ANDROID SMARTPHONES MADE EASY

The Beginners Guide Made For Beginners

By James Bernstein

Contents

Introduction

Smartphones have been around for many years now, and most people are so used to having one that they are just as much a way of life as having a TV or refrigerator in your house. And the younger generation doesn't even know of a life without smartphones, and, between you and me, probably can't function without one!

For those of us that have been around longer than smartphones, you may find it's not as easy getting as comfortable with using one compared to a teenager. Many people today still don't even like to use their home computer and taking on another electronic device can be too much for them. And, worst of all, these smartphones usually don't come with instruction manuals, so you are forced to learn how to use them on your own or with the help of others.

The goal of this book is to get you started on your smartphone adventure without confusing and irritating you at the same time. I find that if you explain things like someone is a total beginner, even if they are not, it makes that topic much easier to understand, and that is the way this book was written—so that *anyone* can make sense of the content without feeling lost.

This book will cover a wide variety of topics such as phone setup and configuration, installing and using apps, texting, making phone calls, using the camera, email and Internet usage, navigation, security, and so on. One thing I will stress over and over throughout this book is that not all Android-based smartphones are alike. There are many upon many manufacturers of smartphones that use the Android operating system (hence the name), and many of them like to use their own interface (or "skin" as it's often called) to customize the way the phone looks and feels to make it unique. For example, a Samsung Android phone will not look and operate exactly the same as an LG Android phone.

For the most part, all Android smartphones do most of the same things, but how you do them is where you will see the differences. So, if you find something in this book that doesn't match exactly with your specific phone, then that is most likely the reason, and you may have to do a little poking around to get the results you are looking for. I will be using a Google Pixel smartphone for my examples in this book, and I like the Google phones because they tend to keep the look and feel pure Android without adding any fancy interface to the phone (which can make things more complicated, in my opinion). This is where iPhone users have the advantage since an iPhone is an iPhone and only Apple makes them and doesn't let anyone else make their own version. (But, in my opinion, it's the *only* advantage!)

So on that note, break out your phone and your reading glasses if you need them, and let's get down to business!

Chapter 1 - What is a Smartphone

Since you are reading this book, it's most likely the case that you own a smartphone or are at least thinking of getting one. If you *don't* have one, then you are in the minority, since smartphones have pretty much replaced land lines for most people and nobody really uses the older cell phones that only make calls and send basic text messages anymore (even though you can still get them).

Think of a smartphone as a phone and computer all in one. It can do the basics like make phone calls and receive voicemails, but it can also do things your computer can do like send and receive email and browse the Internet, among many other things. In fact, many people use their smartphones in place of having a computer since they can do everything they need to do on their smartphone itself. For me, a smartphone is not a replacement for a nice desktop computer with a big monitor, full size keyboard, and a mouse, and will never be as easy to use for most common computing tasks (such as writing this book, for example!).

Smartphone History
If you want to know what the first smartphone model was, then it depends on who you ask, but to me it's when Apple came out with their first iPhone in 2007, and then when Motorola came out with the Android based Droid in 2009 (figure 1.1). T-Mobile released their G1 phone that used the Android operating system in 2008 (figure 1.2), but it wasn't until the Droid came out that things really took off.

Figure 1.1

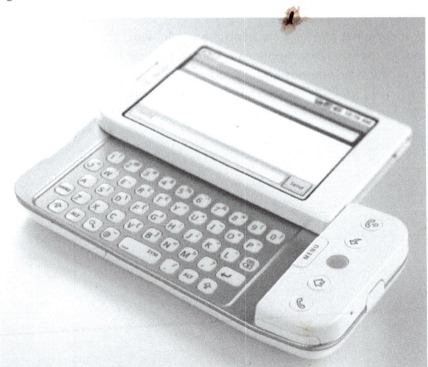

Figure 1.2

These original Android smartphones had many of the same capabilities that current smartphones have such as touch screens, email, and web access, navigation, texting, and built in cameras for photos and videos. Over the years, these phones have just been getting faster and more powerful, with new features being added on top of the existing features.

Going back a little further to 2005, Internet giant Google bought a company called Android Inc. (for around $50 million) who had created an operating system that was meant to be used for digital cameras. Google took the Android operating system and built upon it so it could be used with mobile devices, and the rest is history!

Android Smartphone Brands
There are many companies out there that are licensed to use the Android operating system for the phones that they create, which leads there to be many brands of Android-based smartphones on the market. It's hard to say if this is a good thing or a bad thing since on one hand it gives you many choices, but on the other hand it makes for a mess when it comes to standardization of features and operating system versions.

 An operating system is the software that communicates with the hardware (the smartphone) and allows you to actually use the phone and install apps into that operating system. For you PC users, Windows is the operating system for your home computer and serves the same purpose. iPhones have their own operating system called IOS.

Some of the more popular makes and models (currently) of Android smartphones that you might have heard of include the Samsung Galaxy, Samsung Galaxy Z Fold, Google Pixel, OnePlus 11, Motorola Razr Plus, and Nothing Phone 2. These models are constantly being updated and replaced with newer models, so things are always changing faster than most people can keep up with when it comes to Android smartphones.

The brand of smartphone you will end up with will most likely be determined by what cell phone provider you are using. Not all cell phone providers carry the same models. For example, Verizon will have phones that Sprint doesn't, and so on. Many manufacturers decide what providers they want to make their smartphones for. If you have ever heard of the term "unlocked phone," this means that the phone is not associated with a specific provider and can be activated on more than just one of these providers assuming the provider will support that particular phone. So it's always a good idea to make sure the phone you want will be supported by your cell phone provider before making any purchases.

Plans and Payments

Buying a smartphone and choosing a plan can be as painful as buying a new car if you don't know what you are looking for and what the available options mean. You will have to decide on things such as how much data you think you will use per month, if you want to buy the phone outright or make payments on it, and whether or not you need unlimited text messages.

The most difficult decision you will need to make is how much data you will be using per month. Smartphones have data plans that only allow you to do things online such as browse the web, stream music, or watch videos, which consumes data as you go. If you use more data than allowed, then you will be charged extra for going over, and that adds up really quickly. And, if you get more data than you need, you will be paying for it whether you use it or not. Some plans allow you to transfer any unused data over to your balance for the next month, giving you a little cushion in case you have a heavy data usage month. Fortunately, you should be able to change your data plan at any time to give you more data or take away data that you don't need to use or pay for, so it might take a little trial and error on your part to see what works best for you. (I will be going over how to check your data usage later in the book.)

If you don't get the free phone that comes with many plans, then you will need to either buy the phone outright or make payments on it. If you choose to make payments on it, then you usually get stuck in a two year contract paying it off. Even if you buy the phone outright, make sure you don't get put in a contract that you don't want to be in unless it benefits you to be in that contract. By the way, the free phones are free for a reason, and if this is going to be your first smartphone experience, then it might get ruined by one of these cheap phones!

Android vs. iPhone

There is most likely some reason you chose to get an Android-based smartphone over an Apple iPhone. It might have been because there was a special deal that included the phone, or a friend or relative convinced you that you would be better off with an Android, etc. Whatever the reason, it appears that you have made the right choice!

Like I mentioned earlier in the book, an iPhone is an iPhone while there are many manufacturers of Android smartphones, and they will differ to some degree as to how they look and how they function. If you are Internet savvy, then you can usually find the answer you are looking for online, and once you get better at using your phone and figuring out how to do things, you can eventually figure out how to do almost anything you need to do without wanting to throw your phone out the window!

I always describe iPhones as smartphones for people who just want to be told what to do and how to do it, while Android smartphones are for those who like to have control over their phones and tweak them to work the way they want them to. But at the same time, iPhones tend to be a little more secure since they are locked down by Apple and they don't let anyone tweak anything that shouldn't be allowed to do so. iPhones also tend to be more expensive and are one of the best quality phones you can get when it comes to how they are built. There are many great quality Android phones as well. Apple does a good job at making the operating system for their iPhone work similarly to their iPad tablets, as well as their desktop computers and laptops.

Figure 1.3 shows an example of a typical iPhone, figure 1.4 shows a Samsung Galaxy, figure 1.5 shows a Google Pixel, and figure 1.6 shows a Motorola Razr Plus. Overall, the Android smartphones look very similar, and you might even think the iPhone looks similar to them as well (which in fact, it does a little).

Figure 1.3

Figure 1.4

Figure 1.5

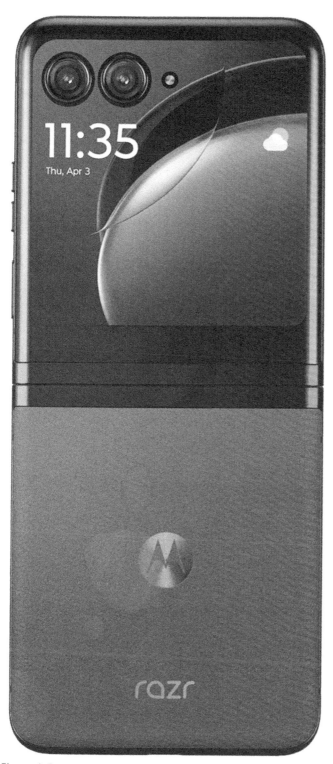

Figure 1.6

Overall, you will find that all Android smartphones have a similar look and feel, but it's mostly how you access features and settings where things start to differ between the models. If you can get proficient on an Android smartphone, then you should be able to get by on an iPhone if you need to use one someday. It's the same thing for computer users that are really good with Microsoft Windows being able to use a Mac without getting too frustrated.

Chapter 2 – Setting Up Your New Phone

When you get your new smartphone, you will have a couple of options when it comes to setting it up for its initial power on and configuration. You can have the store you bought it from do it for you (most likely for a fee), or you can try and tackle it yourself. If you buy it online, then you will most likely be stuck setting it up yourself unless you take it to your local cell provider store and have them set it up for you. If you are feeling a little techy, I would suggest that you do the setup yourself because it will help you get used to how the phone works, and then you will know firsthand how it has been configured.

Initial Configuration

The first time you power on your new smartphone you will be asked a few questions to help configure your phone for its initial usage. These questions will vary depending on what phone you have, and what version of the Android operating system you are running. Just like with the phones themselves, Google updates its operating system from time to time to add improvements and security features.

Even though your phone should come with the battery at least somewhat charged, you might want to connect it to the wall to charge it for a little bit before starting your phone. Smartphones will come with a wall charger and a detachable USB cable that can be used to connect to your computer for charging, as well as transferring pictures and other files. If you take a look at figure 2.1, you will see that you can use the USB cable to connect to the wall charger, or you can connect the USB end to your computer to charge it that way.

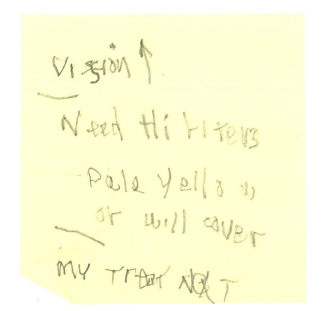

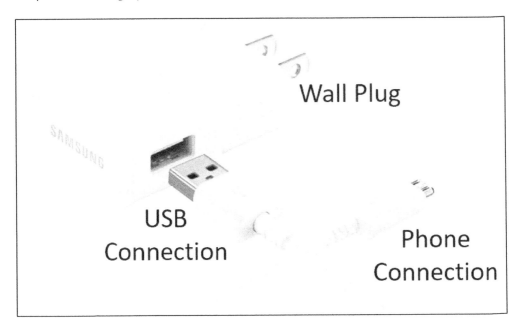

Figure 2.1

To power on your new smartphone, hold down the power button on the side until you see something appear on the screen and then let go. It will take a minute to start up to the point where you will need to start inputting information. Your phone will most likely have one button for power and then two others for volume up and down, and their location will vary depending on your phone model.

Once the phone boots up, you will begin the setup process and be prompted to answer several questions. Here are some possible configuration questions you might have to answer since, once again, this process will vary:

- Selecting your language.
- Activating your phone on your provider's network (this should be an automatic process).
- Connecting your phone to your home wireless (Wi-Fi) network (be sure to have your wireless password on hand if you don't already know it).
- Configuring the fingerprint scanner if you have one (optional).
- Allowing your phone to download any updates.
- Signing into or creating a Google account to be used with your phone.
- Configuring email accounts.
- Setting up security options.

I will be going into more details about things such as the fingerprint scanner and configuring email accounts in later chapters, but want to take a moment to discuss the process of creating a Google account and why you need to do so on Android-based smartphones.

Google Account

There are a few reasons why Google wants you to have a Google account associated with your phone, but the main reason is to keep track of what apps you download from the Play Store, which is the place you search for apps to install on your phone. There are many free apps, and also many that cost money, so having a Google account makes it easy to keep track of what apps you have and which apps you have paid for. You don't need to set up any payment information for apps if you don't want to but will only be able to download free apps this way, which is usually fine for most people (including myself).

Another benefit to having a Google account is that it stores not only your app information, but also your contacts, so that way if you get a new phone, all you need to do is log in with your Google account and all of your contacts will be imported and you are ready to go. You can also go to Google Contacts on your computer and access your contacts from there just by signing in.

Signing up for a Google account also gives you a Gmail account just for doing so. Gmail is Google's email service, and it's free to use and very popular. It's also easy to check your Gmail messages from anywhere you have Internet access. Keep in mind that you don't *need* to use Gmail once you sign up, but it will be there in case you change your mind.

If you already have a Gmail account, then that means you have a Google account, and all you need to do is enter your Gmail address and password during setup and you will be ready to go, and all of your Google contacts and email will be imported into your new phone. You don't have to use your existing Google account and can create a new one during the setup if you choose to do so. It is also possible to have multiple Gmail accounts on one phone if that's something you wish to do.

Setting Up Your Wireless Connection

I mentioned that you will be prompted to configure your wireless Internet connection during the setup of your phone, but if you are not at home when you set up your phone, you will not be able to do this and will have to skip this step. Or, if for some reason something changes with your wireless connection, this will apply to reconfiguring Wi-Fi to work with your phone.

To configure your wireless connection, you will need to find your phone's settings icon, which usually looks like an image of a gear. On my phone, if I swipe down from the notification area (discussed later) at the top of the phone, I can see that my phone is using the cellular connection and not the Wi-Fi (figure 2.2). This is normal if you are out and about and not around a Wi-Fi hotspot or at home. But if you are at home and see this, that means you are not connected to your wireless connection. At the bottom right of the screen, I also have the settings gear icon that I can tap on to bring me to my phone's settings, where I will then tap on *Network & internet* (yours might be named slightly different).

Figure 2.2

Figure 2.3 shows that my Wi-Fi is enabled, and below that I see all of the wireless connections that are in range of my phone. I am going to connect to the wireless connection named McCheese by tapping on it. Next, I will need to type in the password (as shown in figure 2.4) and tap on *Connect*.

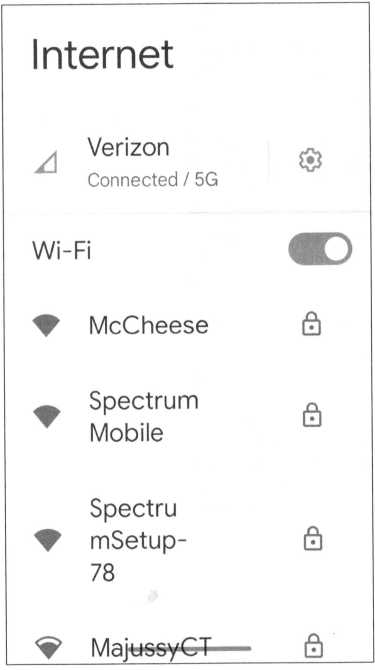

Figure 2.3

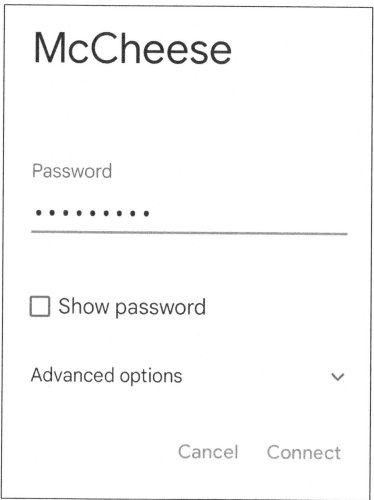

Figure 2.4

Figure 2.5 shows that I am now connected to the McCheese Wi-Fi connection.

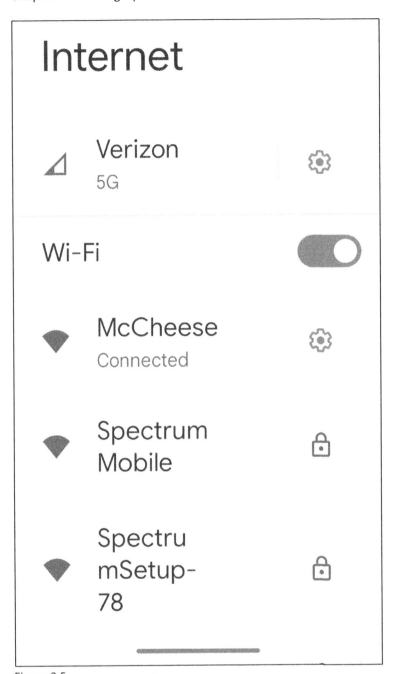

Figure 2.5

If I tap on the McCheese connection name, I can see additional information about that connection such as the signal strength, its frequency, and the security type it uses.

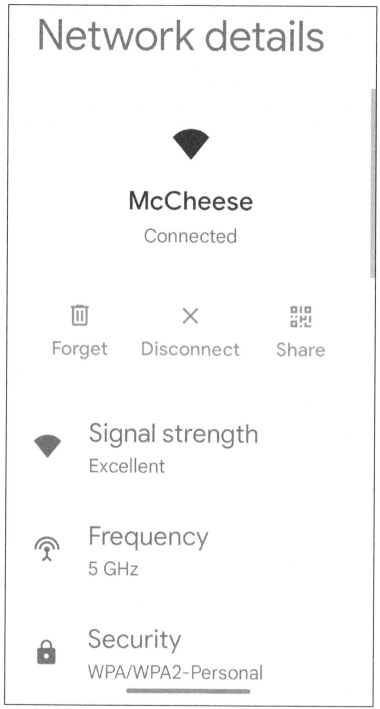

Figure 2.6

Now that I have my wireless Internet connection configured, I can do things like browse the Internet, check my email, and download apps. If you go somewhere like a hotel or friend's house and want to use their wireless connection, you will need to connect to it the first time and enter their password. Then next time you go to that location, your phone will automatically reconnect to that wireless connection (unless the password has been changed).

 It's always a good idea to connect to your wireless when at home so you are not using your cellular data for your Internet activities since you have unlimited data with your home wireless connection. The same goes for when you are at friend's houses, or staying at hotels, etc.

Screen Protectors and Cases

You might have noticed that your new smartphone feels a little on the delicate side, and maybe even a little slippery when you are holding it. These phone manufacturers focus so much on making their products look sleek that they tend to not think so much about making them durable.

When it comes to protecting your phone, you should defiantly get a case for it to help keep it safe from scratches, from dropping it, or even setting it down on a rough surface. There are many types of cases to choose from, ranging from super slim to super rugged. I don't recommend getting one from the store you got the phone from unless you like to over-pay for things. You can go online somewhere like Amazon and get a decent case for $10. Just make sure you are getting one for your exact phone. For example, don't just get a Samsung Galaxy case and assume it will fit on any version but rather search for a Samsung Galaxy 21 case or whatever your model might be. Figure 2.7 shows a few examples of the types of cases you can get for your phone.

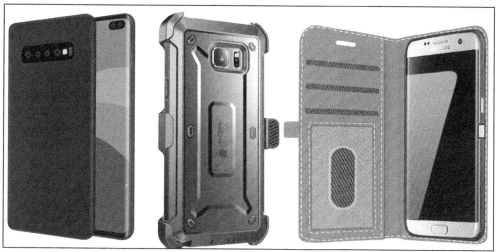

Figure 2.7

Another protective item you might want to look into is getting a screen protector, which will protect your screen from scratches and potential cracks. There are two main kinds you can get for your phone. The first is a plastic film that sticks on the glass and offers protection mainly for scratches. The next type of screen protector is made out of tempered glass and offers more protection when it comes to cracking your screen if you drop your phone. The idea is that the screen protector itself will crack rather than your actual screen. When this happens, all you need to do is take it off and put on a new one. Once again, you can get these online much cheaper than you can at the phone store, and once again make sure you get one that is made for your specific phone, otherwise it won't fit.

Figure 2.8

Chapter 3 – The Home Screen

Now that you hopefully have your new smartphone set up and protected, it's time to start making it a little more user friendly. Many smartphones will come with a bunch of preinstalled apps that you will never use and can't uninstall if you don't want them on your phone anymore. One thing I like about the Google Pixel phones is that they don't come with most of the unnecessary nonsense apps installed like other brands do.

The best way to streamline your phone so it's easier to use is to organize your home screen and only have the apps and widgets (discussed later) there that you actually use. The home screen is similar to your desktop on your computer\PC. Smartphones will have multiple screens that you can access by swiping, and you can have different apps on different screens, making it easy to organize things.

Figure 3.1 shows the main home screen on the left, and then the next screen over on the right. Notice how the main icons on the bottom stay the same for both screens? This is common because here is where you will have your most commonly used apps, and you will usually see things like your phone, contacts, and text messaging app here. Depending on your, phone you may have up to five screens that you can scroll back and forth on and have specific apps and widgets on each one.

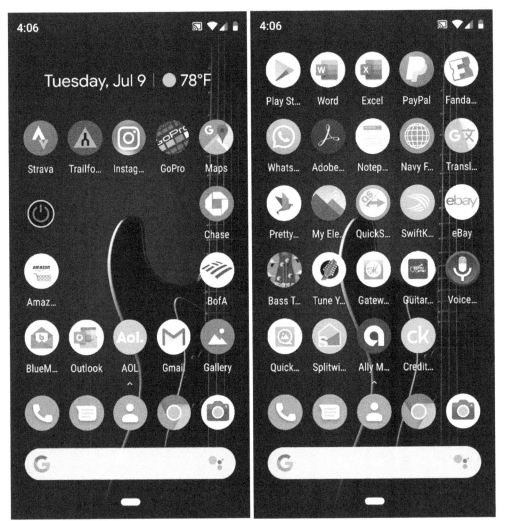

Figure 3.1

One important thing to keep in mind is that these home screens will not show *all* of your installed apps, and the way to see all of your installed apps will vary from phone to phone. For example, you might need to swipe up from the bottom or side of the phone or to see all of the installed apps (like shown in figure 3.2).

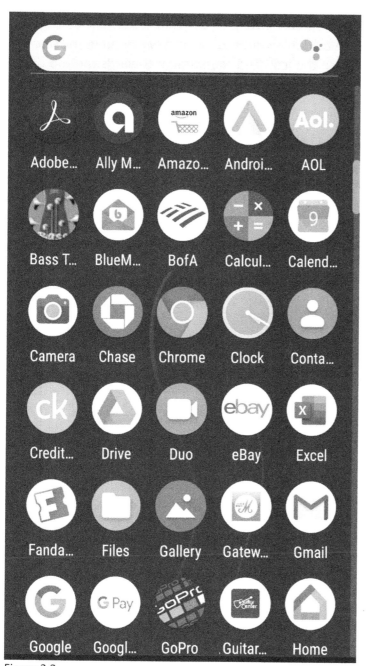

Figure 3.2

Adding and Arranging Your App Icons

Many times when you install a new app it will automatically be added to your home screen at the next available free space. Most people like to have their favorite or most commonly used apps on their main home screen so they don't have to go searching for them when it comes time to open one up.

There are a couple of ways you can arrange your app icons to clean up the appearance of your phone. If you don't want an icon on any of your home screens, you can simply tap and hold on the icon and then drag it towards the top of the screen and either tap on *remove,* or put it in the trash can, depending on the option your phone gives you.

Figure 3.3 shows that there is an option to remove the icon or to uninstall the app itself. Keep in mind that removing an icon from your home screen does not uninstall the app, and you can still get to that app from your main app screen if you decide you want to use it. You can also drag the app icon from your main app screen back to your home screen if you want it back. You simply use the same tap, hold, and drag method you used to remove it.

Figure 3.3

You can also use the tap, hold, and drag method to rearrange your icons and drag and drop them from one home screen to the other. The same thing applies to the apps at the very bottom that stay the same no matter what home screen you are on. These can be rearranged and changed as well by using the same method.

Widgets

Android smartphones allow you to use what they call widgets on your home screens that offer additional functionality to your phone. You can think of widgets as live content that you can view and interact with right from your home screen. To access your available widgets, press and hold on a blank area of your home screen and choose the widgets option. Then you will see your available widgets, and it should look similar to (but not exactly like) figure 3.4.

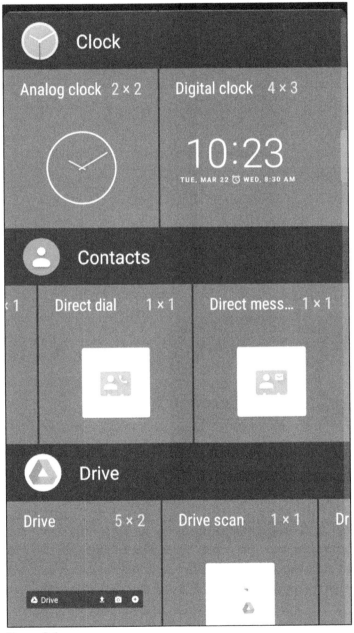

Figure 3.4

The widgets you have available to you will vary depending on what phone model you have and what apps you have installed since many apps will include widgets that you can use on your home screen. For example, if you installed a weather app, it will most likely come with a widget that allows you to see the current weather right on your home screen without having to open the app itself.

I am going to add a direct message widget to my screen for my example. This option will allow me to place an icon on my home screen that will allow me to send a text message to a specific person without having to open my text messaging app and find that person to send a message to. Once I choose the person from my contacts, they are shown on my home screen (as seen in figure 3.5).

Figure 3.5

You might have noticed that this widget looks a lot like an app icon, and some will look like this while others will take up more space on your screen depending on what widget you are using. For example, figure 3.6 shows all the possible size widgets you can use for a particular weather app, and obviously you would choose just one.

Figure 3.6

Changing Your Wallpaper

Just like with your desktop computer (and laptop), Android smartphones will allow you to change your background wallpaper to suit your style. Most phones will have built-in wallpapers that you can choose from, or you can use a picture that you have taken with your phone or downloaded off the Internet as your background.

As you can see in figure 3.7, my current background image is a guitar against a black background. To change my wallpaper image all I need to do is long press on a blank spot of my home screen and choose *Wallpaper & style* to get to my wallpaper options. Once again, this process may vary a little on your particular device.

Figure 3.7

Then I will be shown my available options for what pictures I have available to set as my wallpaper. Figure 3.8 shows my current wallpaper and some other options for color themes and additional pictures. If I scroll down a little, I can tap on *More wallpapers* to see other picture options I have (as shown in figure 3.9).

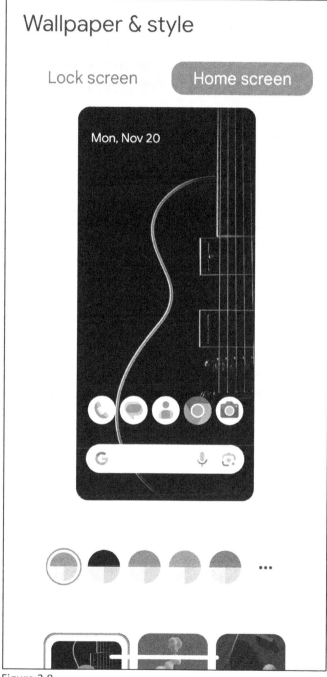

Figure 3.8

Figure 3.9

If I choose the *Landscapes* category (figure 3.10), I will be shown all the landscape-type pictures that come included with the phone.

Figure 3.10

I will then choose the first picture from the Landscapes category by tapping on it. Then will have an option to preview the picture or set it as my wallpaper. I will also have the option to set it for my lock screen and my home screen.

If you choose home screen, it will only show on your home screen while your phone is unlocked. If you choose lock screen, then it will appear on the screen you see when your phone screen is on, but the phone is still locked. Then, of course, you can choose the option to have the wallpaper applied to both.

Figure 3.11

You might have the option to view information about this picture by tapping on the i in the circle at the upper right of the image.

Figure 3.12

Figure 3.13 shows what my phone looks like with the new wallpaper applied.

Figure 3.13

I always like to play around with wallpaper pictures because some look better than others. It's also a good idea to choose pictures that are tall rather than wide to match the orientation of your phone. There are websites you can go to that offer high resolution phone wallpapers on just about any topic you can imagine.

Chapter 4 – Notifications

With today's fast-paced lifestyle, people tend to want to know everything as it happens and nobody has the patience to wait until they get home to see what they missed. Thanks to smartphone notifications, we can be informed of things we need to know about (or at least think we need to know about) as they happen.

For example, if you are expecting an important email, you can have your phone notify you when new email arrives rather than have to open up your phone and email app and check for yourself. The same thing applies to other applications like text messages, banking apps, shopping apps, and so on.

Fortunately, these notifications are customizable so you don't get stuck hearing constant notifications all day long for things you don't care about. Generally, you can turn notifications on or off on a per-app basis, as well as customize how you get notified in regard to sounds, vibrations, visual cues, and so on.

Notification Bar

All of your notifications will show up in what is called the notification bar. This notification bar is usually at the very top of your phone and will show certain icons for various types of notifications. Then you can view these notifications by pulling down with your finger from the top of the bar.

Figure 4.1 shows notifications for the Gmail and Outlook email apps at the top of the phone. Also notice that there is a small notification dot on the app icon itself. Not all apps have this feature, but you will find that many do.

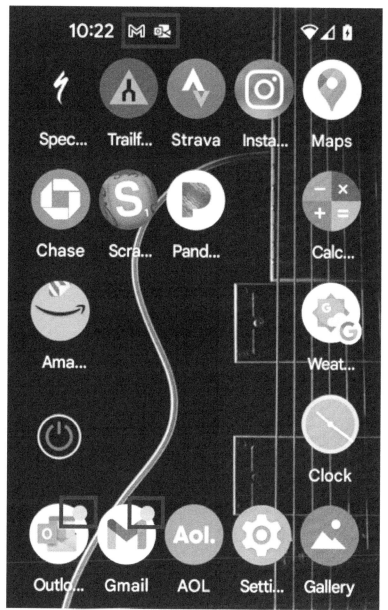

Figure 4.1

When I use my finger and pull down from the top notification bar, I am shown more details about the notifications themselves (as shown in figure 4.2).

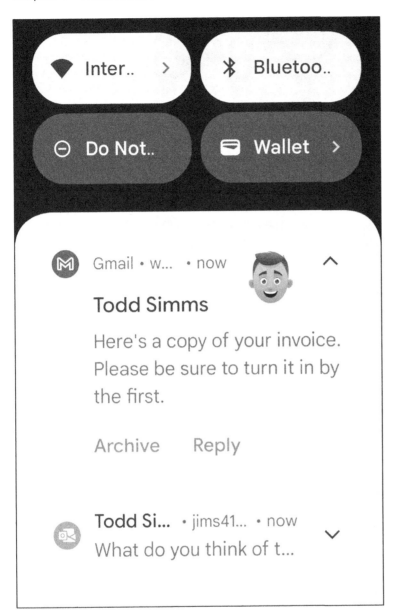

Figure 4.2

From here I can tap on any one of the notifications to open it up and read the email (in this example). At the bottom of the screen, you will see there is an option that says *Clear all*. If you tap this, it will clear all of your notifications out, but will not do anything like delete the email or text messages etc. that it was referring to. You can also swipe a particular notification to the right to clear just that one notification message. There is also the option to manage notifications, which I will be going over later in this chapter.

You should also be able to see notifications on your lock screen so that when you turn your screen on, you can see these notifications without having to unlock your phone (figure 4.3). Many phones will let you open a notification right from here.

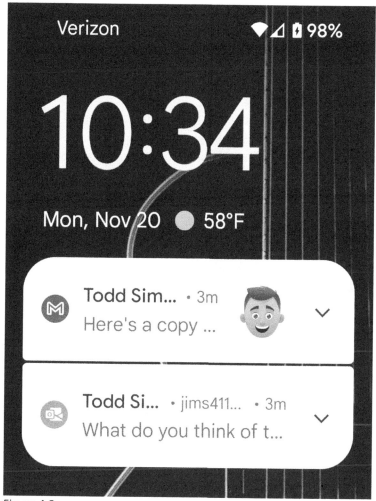

Figure 4.3

Managing Notifications

If you want to control how your notifications notify you, then that is possible by going to your notification settings, OR by going to your apps and then going to the notification options from there. Once again, this process will vary from device to device.

When you pull down to see your notifications, you may have an option to tap on that says *manage notifications* and it will take you to your notification settings. If you don't have this, then look for your Settings app and then find the notifications setting in there. It might not be on your home screen, but rather in your main app section.

Figure 4.4

Once you are in your settings you should look for something called *notifications* or *apps and notifications*.

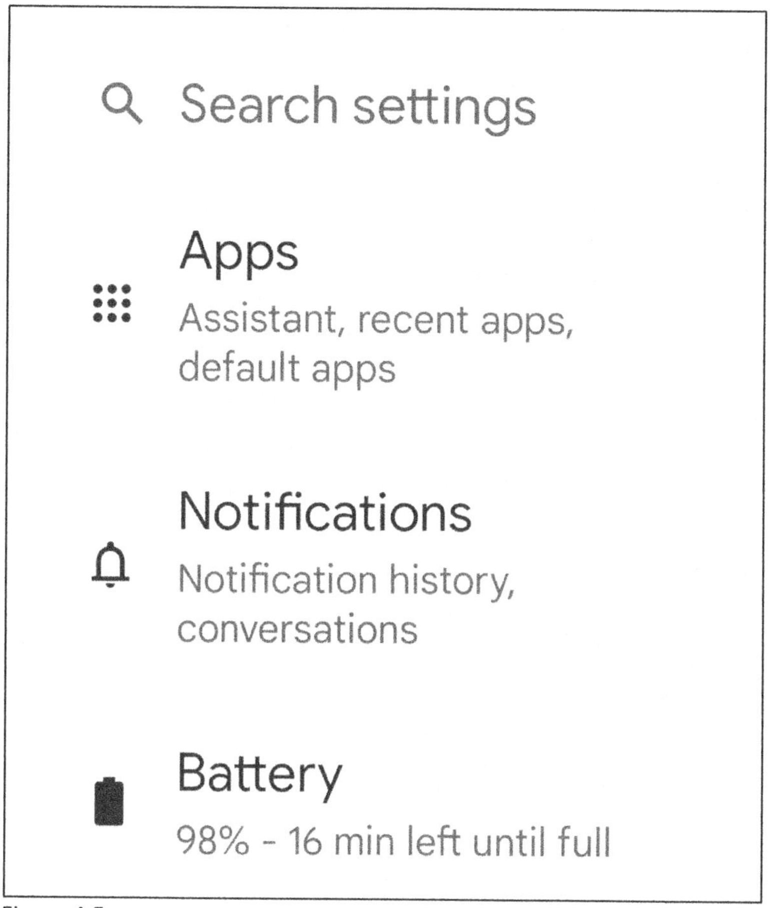

Q Search settings

Apps

Assistant, recent apps,
default apps

Notifications
Notification history,
conversations

Battery
98% - 16 min left until full

Figure 4.5

From there you should be able to see the general settings for notifications that apply to all of your apps (figure 4.6). If you don't want to deal with notifications at all, then you can disable them. There should also be an option for the default notification sound that your phone uses. It's possible to have different sounds for different apps, but generally you set this up within the app settings itself. If your phone has a notification LED light, then you can enable this as well so when you have a notification you will see a little flashing light on your phone, even when the screen is off, telling you that you have some sort of notification. You may also have settings for lock screen notifications and even a notification history section.

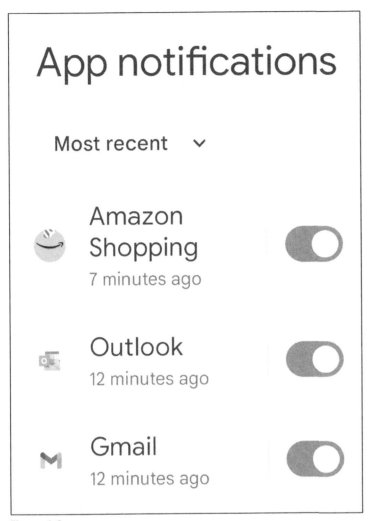

Figure 4.6

Some phones have the ability to show different colored lights for different types of notifications, and you can customize this from the phone settings or sometimes via a third party app that you can download and install.

Depending on the model of your phone, you may have the ability to turn notifications on or off for specific apps all from one place (like shown in figure 4.7). So if you find that you are getting bombarded with notifications from an app that you don't care about, you can turn off the notifications for that particular app.

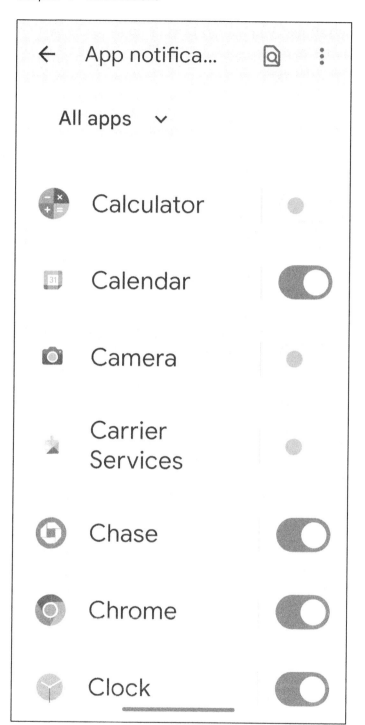

Figure 4.7

If you don't have app-specific settings, you can go to your applications section within your settings, tap on the app you want to modify, and then tap on Notifications to make changes to that app.

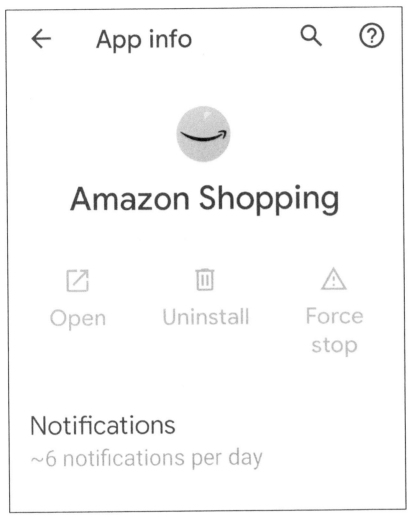

Figure 4.8

Figure 4.9 shows the notification options specific to the Amazon shopping app, while figures 4.10, 4.11, and 4.12 show options specific to the other apps I have installed. As you can see, different apps have different options for how they can notify you.

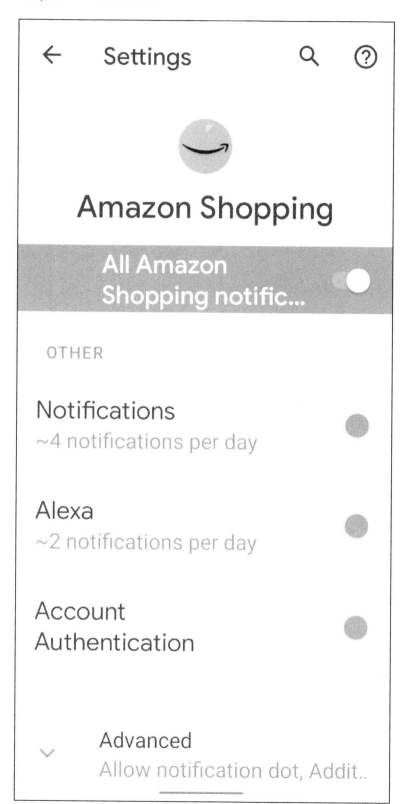

Figure 4.9

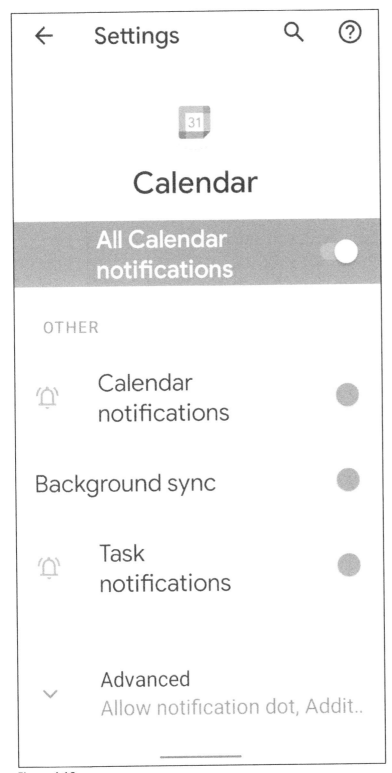

Figure 4.10

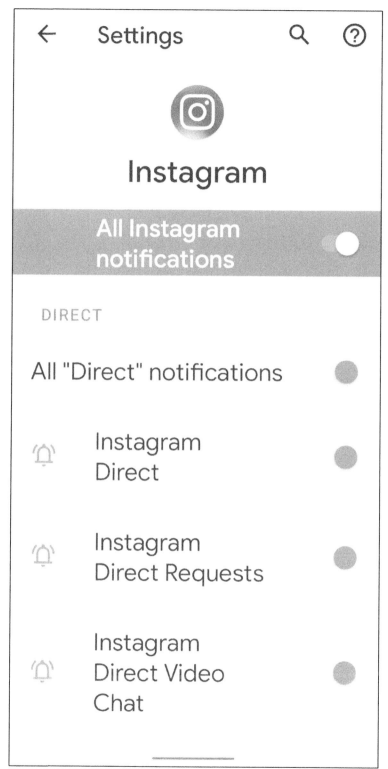

Figure 4.11

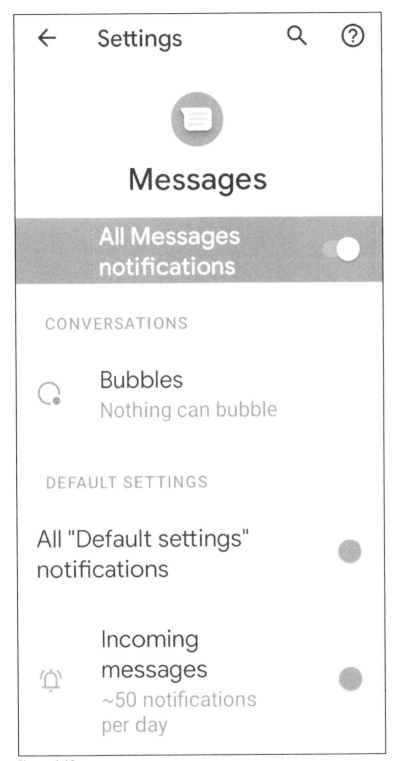

Figure 4.12

One of the best ways to customize your notifications is to assign apps their own notification sound. That way when you hear that particular sound, you know what app it's for and can decide if it's something you need to address now or can just deal with later.

Depending on your phone and Android version, you may or may not be able to easily set notification sounds on an app-by-app basis. When you go into the notification section for an app, look for a section that says Sound (figure 4.13) and you will be taken to a screen where you can choose a different sound for the app's notification (figure 4.14).

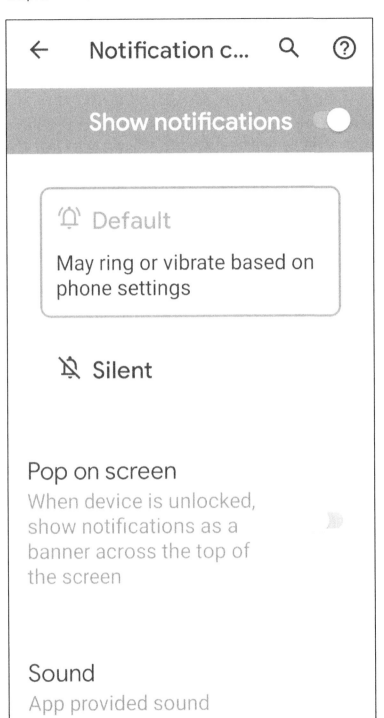

Figure 4.13

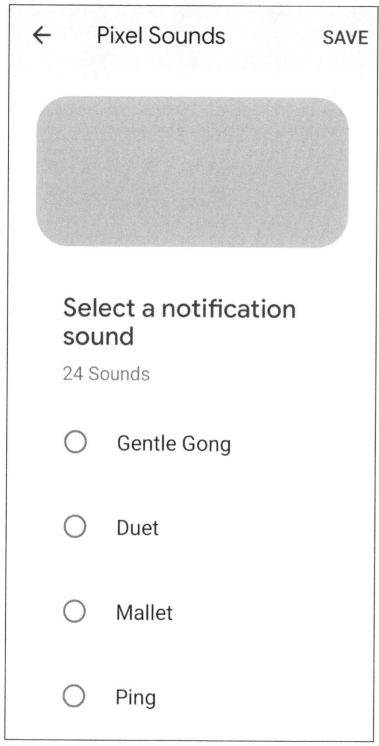

Figure 4.14

From here you can sample other sounds, and when you find one you like you can save it. You can even use custom sounds, but that's beyond the scope of this book.

If you can't change sounds on an app by app basis then you should at least be able to change your overall default notification sound from Settings and then *Sound & vibration*. Many apps will have their own unique notification sound configured when you install them which will help you differentiate between notification types when you hear them.

You are invited to help plant a

BUTTERFLY
WAYSTATION

as part of Cassie Gilbert's Girl Scout Gold Award Project

ARBOR RIDGE PLANTING DAY

- Saturday, July 13, 2024
- 8 AM
- Front Lawn

Arbor Ridge residents are invited to join the

MAGNIFICENT MONARCH CLUB

information will be available on planting day

Chapter 5 – Installing Apps

Without apps, a smartphone is pretty much just a phone with a nice camera, and nobody wants to pay $75\month just to have a nice camera! Your phone will come with many apps preinstalled such as a calendar, web browser, photo gallery, text messenger, and so on. From there, it's up to you to determine what you want to do with your smartphone, and if there's an app out there that will help you do it.

Apps are very similar to the software (or programs) you have installed on your computer. For example, if you use Microsoft Word at home or at the office, then that is software that has been installed on your computer. Apps are software that gets installed on your smartphone, and just like with computers and software, you can have multiple apps installed on your phone.

Google Play Store
In order to install apps on your phone, you need a place to get them from. This is where things differ from computers because you can't insert a CD, DVD, or flash drive into your phone to install these apps. They actually need to be downloaded online before they can be installed.

The place that you go to get apps for your phone is called the *Google Play Store,* and it's actually an app itself that comes installed on your phone. Most of the time this app is placed right on your home screen so you can easily find it.

Figure 5.1

When you open the Play Store app, you will see a screen similar to figure 5.2. On the top you will have various categories such as games, apps, books, movies, and so on. You can also view the most popular apps and view apps in various categories. At the top of the screen there is a search box where you can type in or speak the type of app you are looking for.

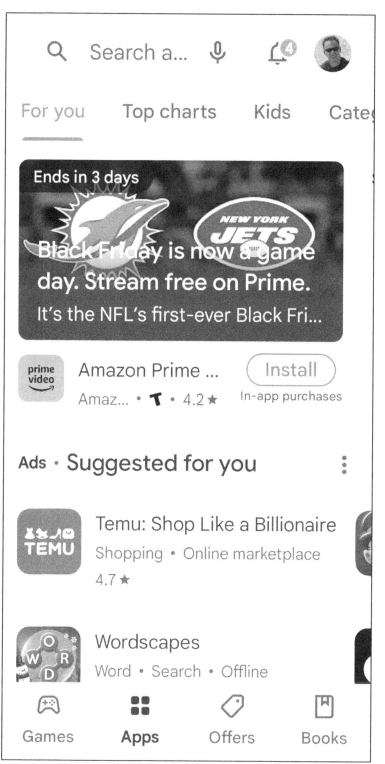

Figure 5.2

For my example, I will look for an app that tracks my steps as I walk, so, therefore, I will type in *step tracker*. Figure 5.3 shows the top results. There will most likely be many pages of results and advertisements for you to scroll through if you choose to take the time to do so. Under the name of the app you will see the developer's name and the rating provided by others who have tried the app and decided to give it a review.

As you can see, even though I typed in step tracker, it was not the first result in my search so make sure you have found the app you are looking for before installing it.

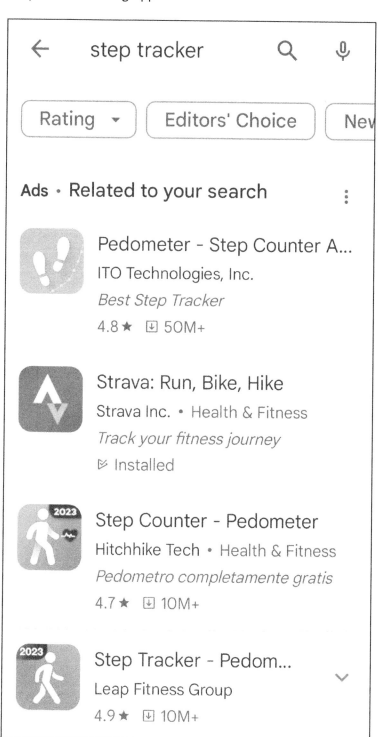

Figure 5.3

I will choose the first app from the search results, and now I get a little more detail when I go into it. It shows me the average review, the number of reviews, and how many times it's been downloaded. I will get into why this information is important later in the chapter.

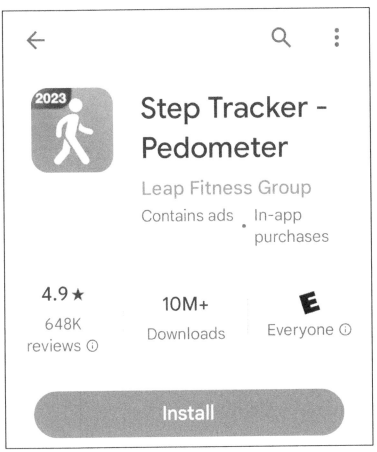

Figure 5.4

If I scroll down further, I will see a little more information about the app, and if I were to tap on the right pointing arrow, I will get even more detailed information about the app.

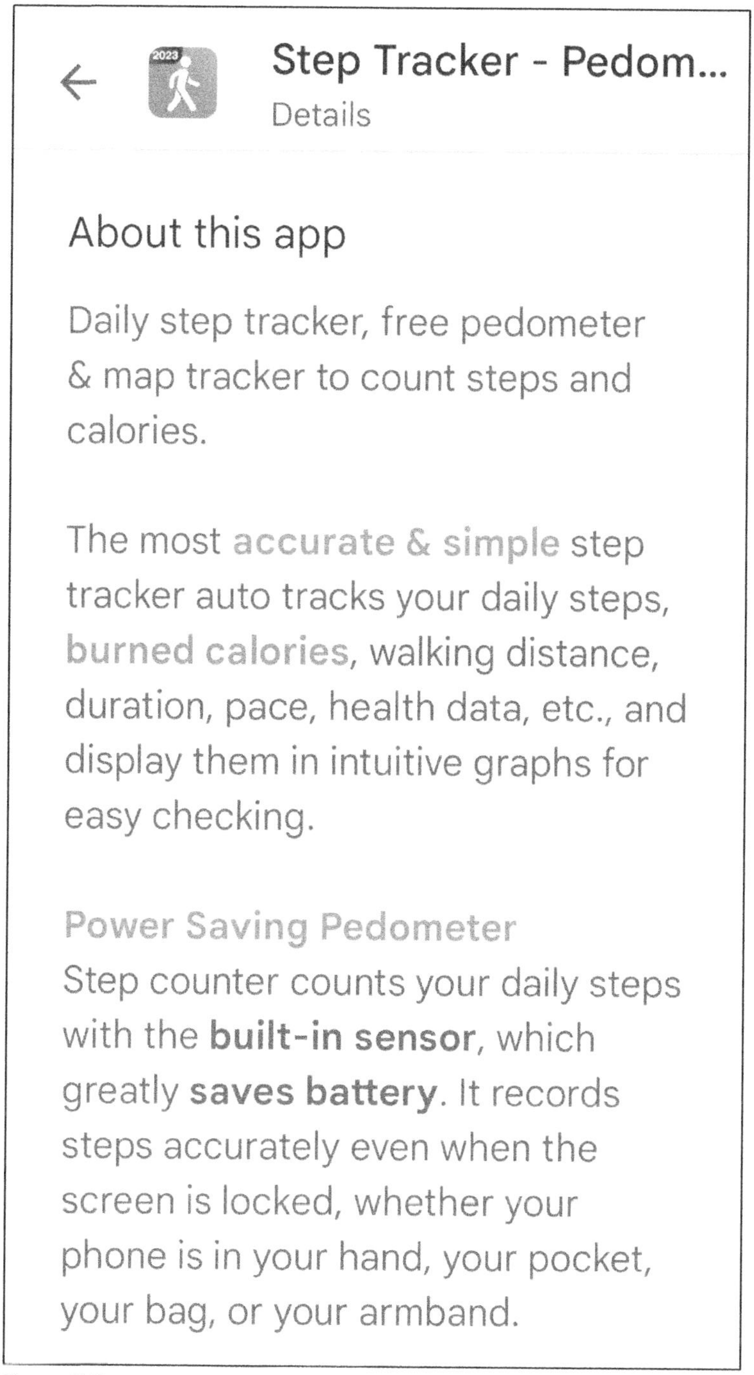

Step Tracker - Pedom...
Details

About this app

Daily step tracker, free pedometer & map tracker to count steps and calories.

The most accurate & simple step tracker auto tracks your daily steps, burned calories, walking distance, duration, pace, health data, etc., and display them in intuitive graphs for easy checking.

Power Saving Pedometer
Step counter counts your daily steps with the **built-in sensor**, which greatly **saves battery**. It records steps accurately even when the screen is locked, whether your phone is in your hand, your pocket, your bag, or your armband.

Figure 5.5

Going back to figure 5.4, if I tap on the *Install* button, my phone will then download the app and then install it (as shown in figure 5.6).

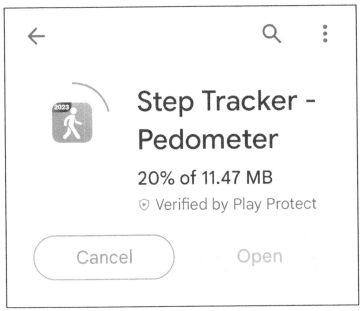

Figure 5.6

Then I will be able to open it right from the Play Store once it's finished installing. It will also place an icon for the app on my home screen. Which screen it places it on will depend on where you have room for another icon and how your phone works.

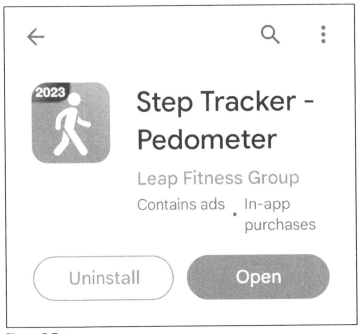

Figure 5.7

There is more to the Play Store than just installing apps (since you can get other things like books and movies), but most people use the Play Store mainly for apps. In fact, in my 10+ years of using smartphones, I have only used the Play Store for apps and nothing else. Maybe I'm missing out?

Free vs. Paid Apps

Now that you have an idea of how the Play Store works, I want to take a moment to discuss free vs. pay-for apps because there is a reason to buy apps even though you can find pretty much anything you need for free. In fact, when you search for an app, you will usually find many more free apps than you do apps you need to pay for.

It's easy to tell when an app is not free because it will show the price of that app next to its name on your search results. You can even have the Play Store show you the most popular paid apps that are being downloaded by others. Notice in figure 5.8 that there is a price next to each one of these apps.

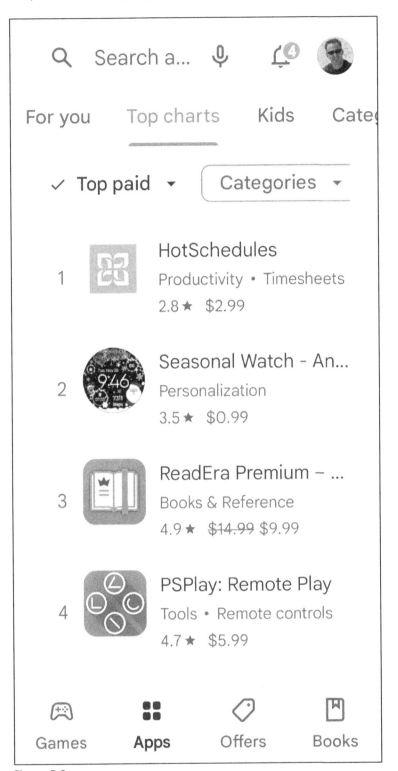

Figure 5.8

When I tap on one of these pay-for apps I don't have an Install button but rather a button with the price shown on it.

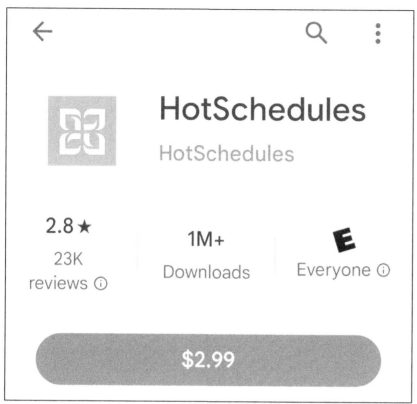

Figure 5.9

When I tap on the price it takes me to a screen with my payment method and a button that says *Buy* that I can tap on to complete my purchase. Then it will install the app just like you saw in my previous example. As you can see in figure 5.10, I have a Visa credit card associated with my account that I can use in the Play Store for purchases.

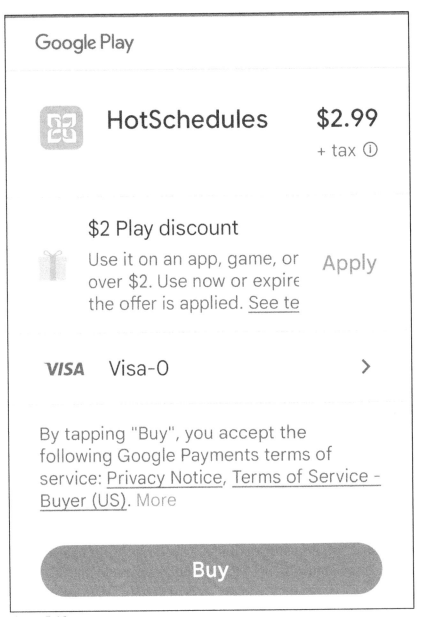

Figure 5.10

If I tap on the right arrow next to the credit card, I will be given a chance to enter a different payment method such as using PayPal or the option to have the price of the app billed to my wireless provider account if that's something my account will allow. As you can see in figure 5.11, either my account is not set up for this, or it's not allowed by Verizon.

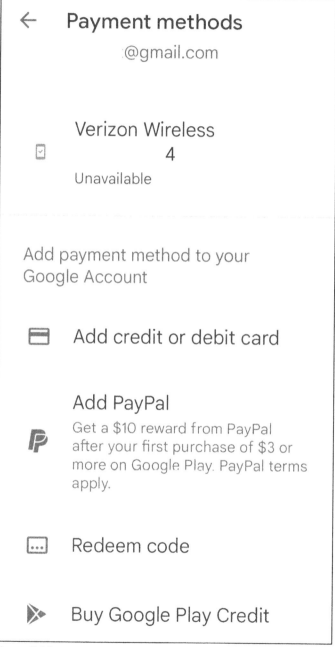

Figure 5.11

There are a few reasons you might want to buy an app rather than use a free one, and I will now go over the top three reasons for buying an app. The first one would be if that's the only version of the app available. If you find an app that appears to do everything you want it to do compared to the free ones you have checked out, then buying it would be the way to go.

The next reason for buying an app would be for its additional features over the free version. Many apps will have a free version and a pay-for version with the pay-for version having more bells and whistles. They let you use a stripped down version for free to try and entice you to buy the full version to get all of the features enabled.

The final reason that I think buying an app is the way to go is if you are not the type that likes to be bombarded with advertisements. Most of these free apps are free because they make money by showing you ads within the app while you are using it. They can be anything from a small banner ad at the bottom of the screen to forcing you to watch a random 30 second video whenever the app sees fit. Some apps are much worse than others when it comes to their ads getting in the way, with games being the worst offenders.

Checking Reviews
Before you download any apps to your phone, especially free ones, you should always check them out first to make sure that they are what you are really looking for, and that they are not something that will cause you more trouble than it's worth.

One major difference between Androids and iPhones is that Apple does a much better job of controlling what apps can be installed on iPhones compared to Google with their Android phones. With that in mind, don't go thinking the Play Store is a dangerous place with cybercriminals waiting to steal your information, but just keep in mind that there is less control over what apps show up there compared to Apple's App Store.

What I like to do before installing a new app is read through some of the reviews, especially the bad ones or the middle of the road (3 star) reviews to see what others think of the app and how their experiences went. When someone puts out an app in the Play Store, it's easy for them to have all their friends download the app and give it a 5 star review, so you can't always trust those.

Going back to the step tracker app I installed in my example, if I scroll down to reviews and then tap on the average reviews, then I can then sort the reviews by things like positive, critical, number of stars, and so on. Since I want to see the worst reviews the app received, I will choose *Critical*. Then I can scroll down and read the awful things people had to say about the app and see if I think they are bad enough for me not to want to use the app.

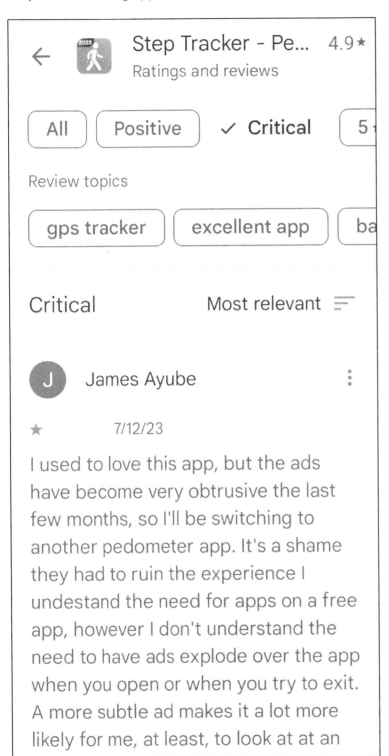

Figure 5.12

If you have an opinion on an app that you want to share, then you can even write your own review for an app that you have installed by going to the Play Store, finding that app, and then giving it a rating.

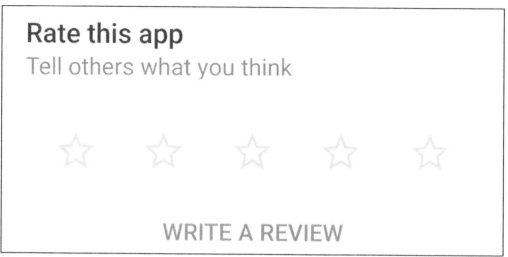

Rate this app
Tell others what you think

☆   ☆   ☆   ☆   ☆

WRITE A REVIEW

Figure 5.13

Checking Download History

Another thing I like to look at before downloading an app is how many times it has been downloaded previously. If there have been hundreds of thousands or even millions of downloads, then I generally feel better because that shows that a lot of people are using the app and it's been tried and tested. If I see something with only a few hundred reviews, then that tells me either nobody is wanting to try out the app, or that it's very new and maybe I should not make myself the guinea pig by being one of the first people to try it out.

So, if I choose another one of the step counting apps from my search, I can see that this one has only been downloaded around 1,000 times compared to the app I installed, which has over 10 million downloads. Notice how it only has 16 reviews and a rating of 3.7. I like to stick with apps that have an average rating of at least 4 or higher.

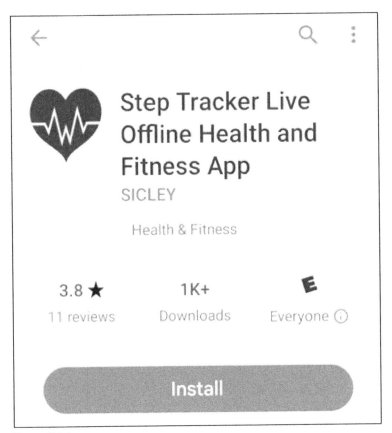

Figure 5.14

Uninstalling Apps

Just because you install an app on your smartphone doesn't mean that you are stuck with it. When it comes to installed apps, you have the ability to uninstall them just like you can uninstall software on your computer. This is a very easy process to do and is actually easier than uninstalling software on a computer.

There is more than one way to uninstall an app from your phone, and I will first go over the most universal method that should apply to any model of Android smartphone. If you open your phone settings by either finding your settings icon or pulling down from the top of the screen and tapping on the settings gear icon, you will then see a choice called Apps where you will find all of your installed apps.

From there, you should be able to see all your apps listed in alphabetical order as seen in figure 5.15. They may be sorted by most recently used and if so, you should have an option to show all your apps.

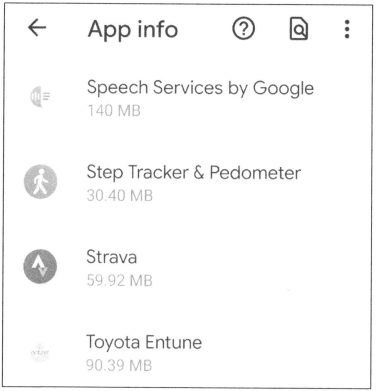

Figure 5.15

I scrolled down to the apps that begin with the letter S so I can show you that I will be uninstalling the Step Tracker & Pedometer app that I recently installed. To do this I can simply tap on the app and then tap on *Uninstall* from the next screen.

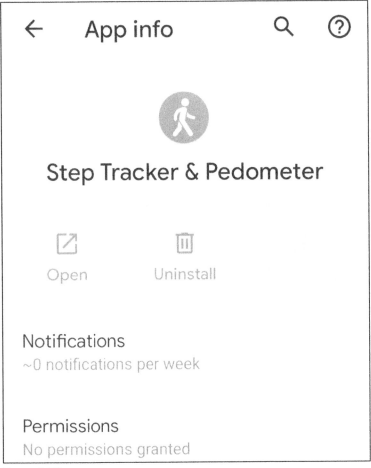

Figure 5.16

Another way to get to your app settings and to also see which apps have available updates is to tap on your profile icon in the Play Store and then tap on *Manage apps & device* (figure 5.18).

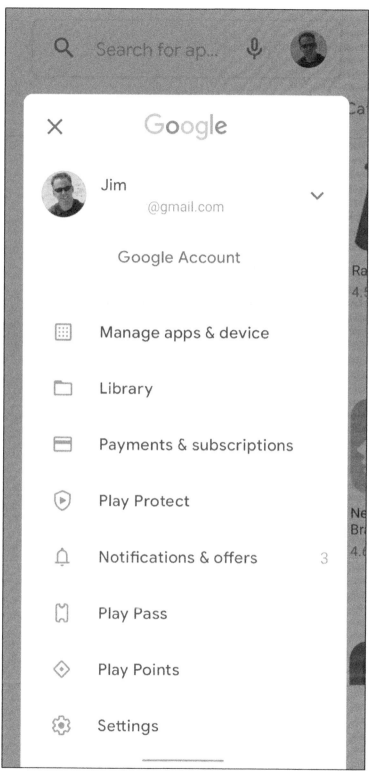

Figure 5.17

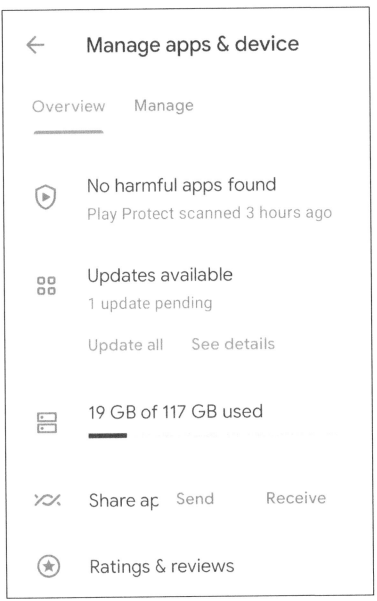

← **Manage apps & device**

Overview   Manage

No harmful apps found
Play Protect scanned 3 hours ago

Updates available
1 update pending

Update all   See details

19 GB of 117 GB used

Share ap Send   Receive

Ratings & reviews

Figure 5.18

From here you can click on *Manage* to see additional details about apps that need to be updated and details on your other apps as well as how much space these apps are taking on your phone.

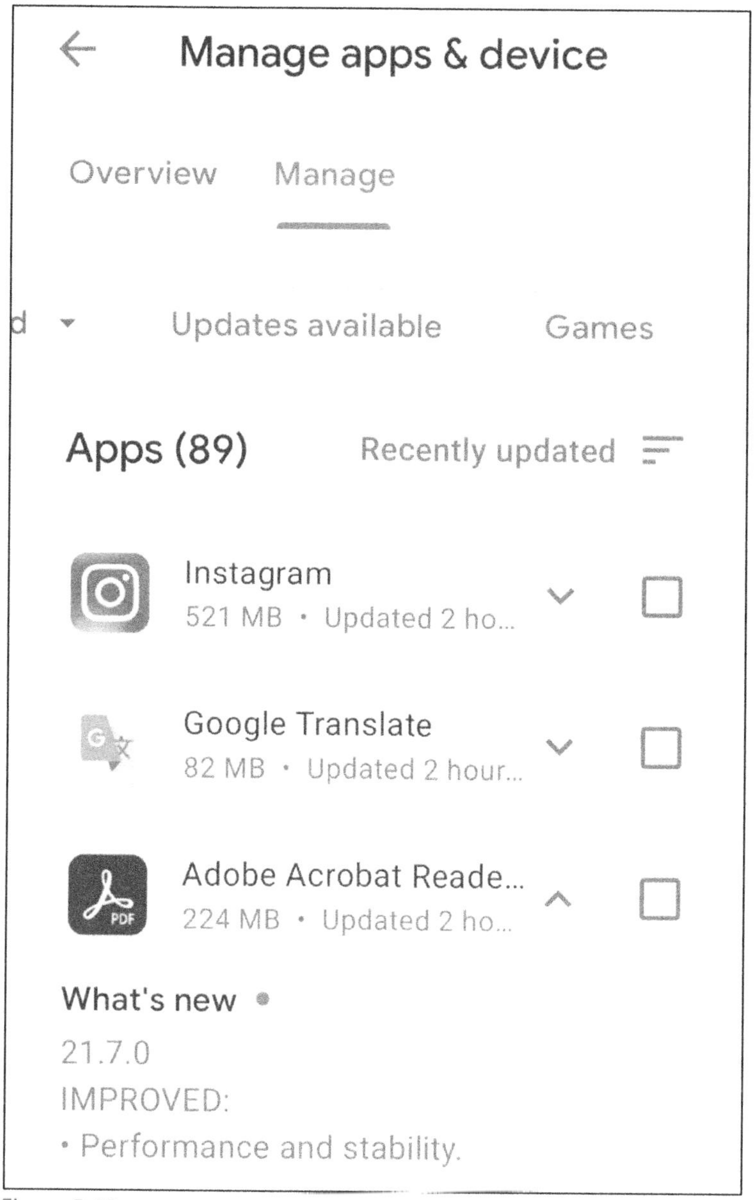

← **Manage apps & device**

Overview   Manage

d ▾   Updates available   Games

Apps (89)   Recently updated ☰

Instagram
521 MB · Updated 2 ho...   ⌄   ☐

Google Translate
82 MB · Updated 2 hour...   ⌄   ☐

Adobe Acrobat Reade...
224 MB · Updated 2 ho...   ⌃   ☐

What's new ●
21.7.0
IMPROVED:
· Performance and stability.

Figure 5.19

As you can see in figure 5.19, I have 89 apps installed, and I have sorted them by recently updated. Taping on *Updates available* will show only the apps that need to be updated.

Now I will uninstall the Step Tracker app that I recently installed by tapping on the app name itself. Then all I need to do is tap on the *Uninstall* button, the app will be removed from my phone, and the icon on my home screen should disappear as well.

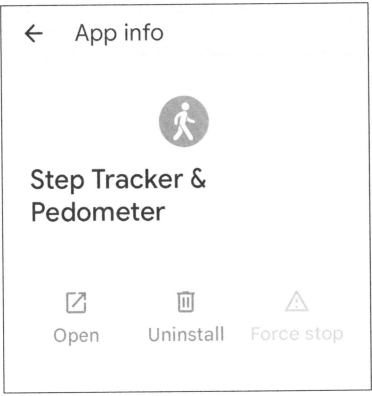

Figure 5.20

Another quick way to uninstall apps is to tap and hold on the app icon and drag it towards the top of the screen on to the word *Uninstall*. Depending on your phone model, you may or may not get an option to Uninstall the app right from the home screen (as seen in figure 5.21). The *Remove* option will only remove the application's icon from your home screen, but not remove the app from the phone itself.

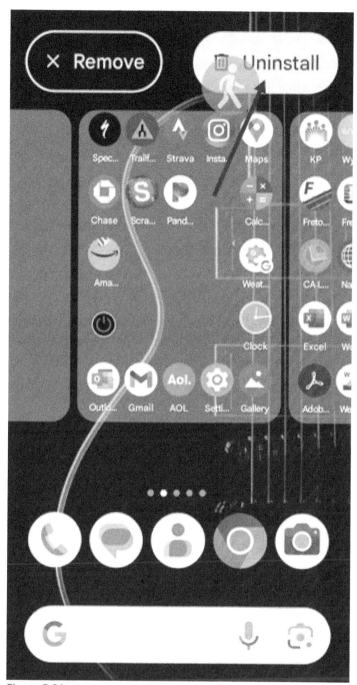

Figure 5.21

Chapter 6 – Setting Up Email

One of the more common things people use their smartphones for is to check their email. You can use your phone to check the same email accounts on your phone that you use at home and even your work email, assuming your company allows you to do so.

There are a couple of ways you can set up your phone to access your email, and I will go over both of them so you can decide which way works best for you. There will also be cases where you can only use one of the two methods that I will be going over if your email provider doesn't support both.

Android's Built-in Email App
Most Android smartphones have a built-in email app (or client, as it's often called) that you can configure to allow you to send and receive emails from your phone and even check your calendar if it's tied into your email account. Remember that your phone will have its own standalone calendar that you can use in addition to or instead of the one that you use with your email. Using this method is more of an advanced way to set up your email, so you might be better off using a vendor-specific email app, which I will be going over in the next section.

To add or configure a new email account on your phone, you will need to find your account settings. This can usually be found under your main phone settings, and if you forgot how to get there, you just need to find your Settings app, which usually looks like a gear icon. From the settings screen, look for an option that says Passwords & accounts or something similar.

Here you will be able to see all of the existing accounts on your phone. Keep in mind that there will be other accounts listed here that are not used for email. For example, as you can see in figure 6.1 there is an Office and WhatsApp account which are not used for email purposes.

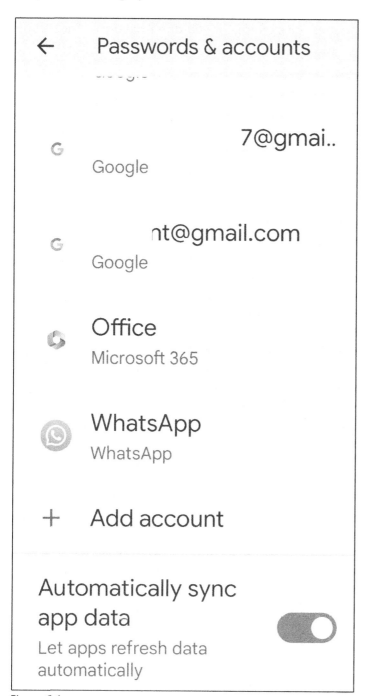

←   Passwords & accounts

7@gmai..
Google

nt@gmail.com
Google

Office
Microsoft 365

WhatsApp
WhatsApp

+   Add account

Automatically sync
app data
Let apps refresh data
automatically

Figure 6.1

To add a new email, all you need to do is click on *Add account* and you will be presented with options similar to (but not exactly like) figure 6.2. As you can see, there are many options to choose from, and there are even more than what is shown in the example. The choices that you will have will depend on what apps you have installed on your phone. For example, if you don't have the Adobe app installed, then you won't have the option to create a new Adobe account.

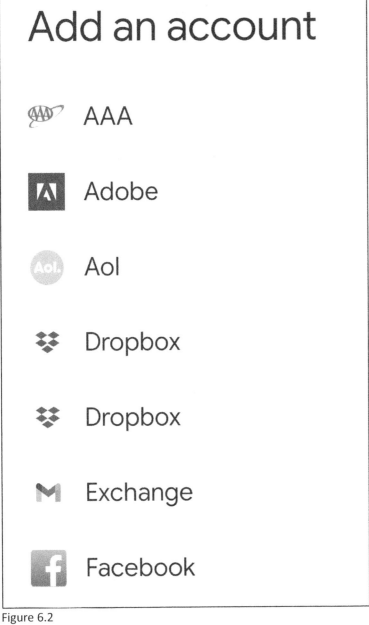

Figure 6.2

If you are going to configure an email account that is not one of the built-in types that there are apps for such as Gmail or Outlook, you will first need to know if your account is an IMAP or POP3 account, so you might have to check with your email provider and get that information and any additional settings unique to their system. For the most, part your phone should be able to contact the email provider and configure itself, so first you will enter in the email address you use for this account and click on *Next*.

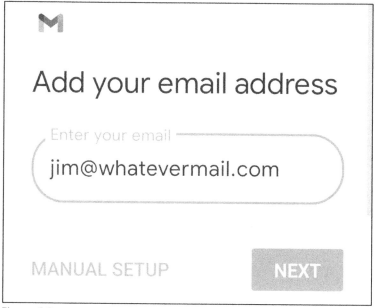

Figure 6.3

Notice how there is an option for doing a manual setup if you want to enter all the email server information in there yourself. Once again, this is more of an advanced way to set up your email account.

Then your phone will try and validate your email account with the email provider's servers. If everything works out, you will just be asked for your password, and then you will be ready to go.

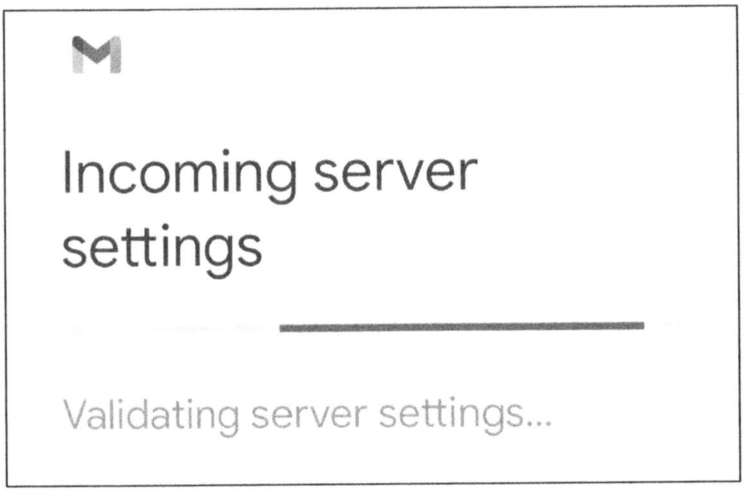

Incoming server
settings

Validating server settings...

Figure 6.4

If it has trouble finding your account or needs you to enter in additional settings, you will be prompted to do so. This is where you would enter the server settings that you obtained from your email provider.

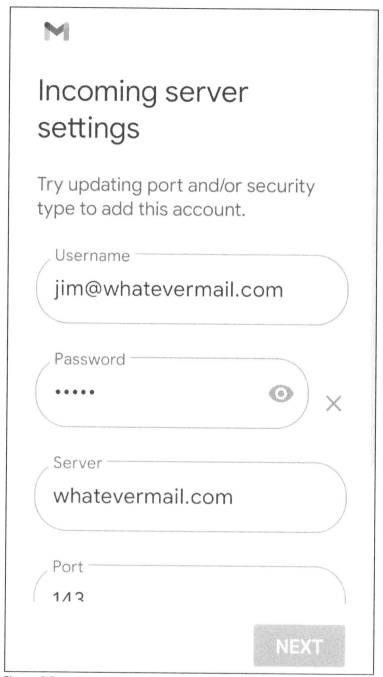

Figure 6.5

If everything works out, then you will be asked to configure a few options if you want to change anything (figure 6.6), then your account will be ready to go. You can find it in the Gmail app, along with any other accounts you have configured (figure 6.7).

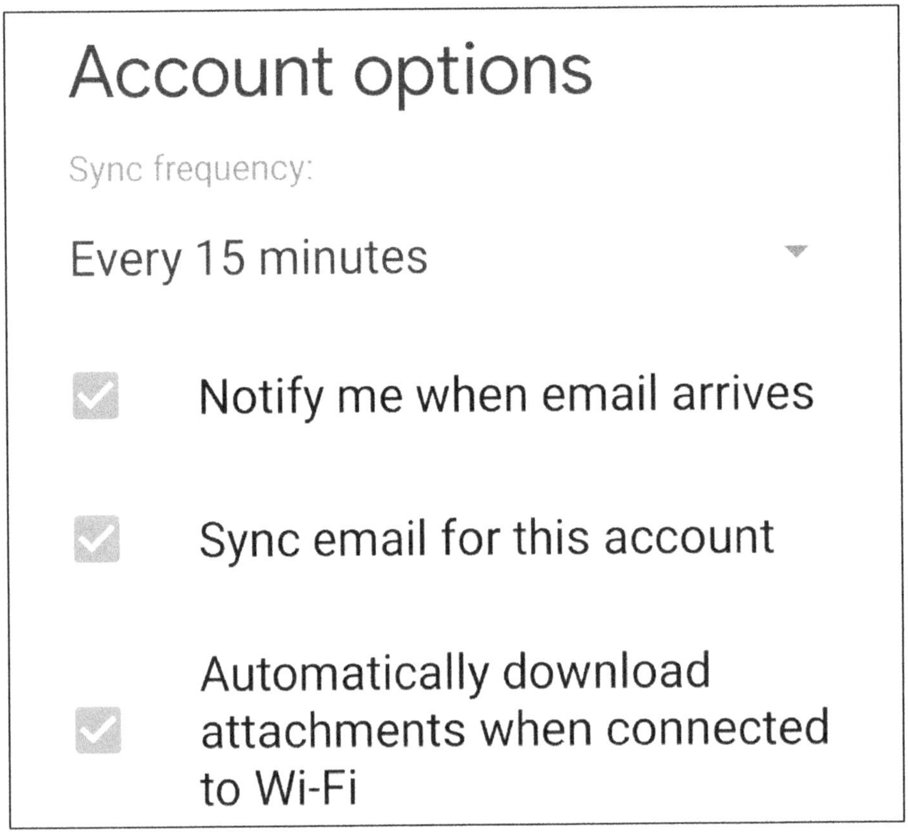

Account options

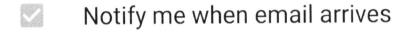
Sync frequency:

Every 15 minutes ▼

☑   Notify me when email arrives

☑   Sync email for this account

☑   Automatically download attachments when connected to Wi-Fi

Figure 6.6

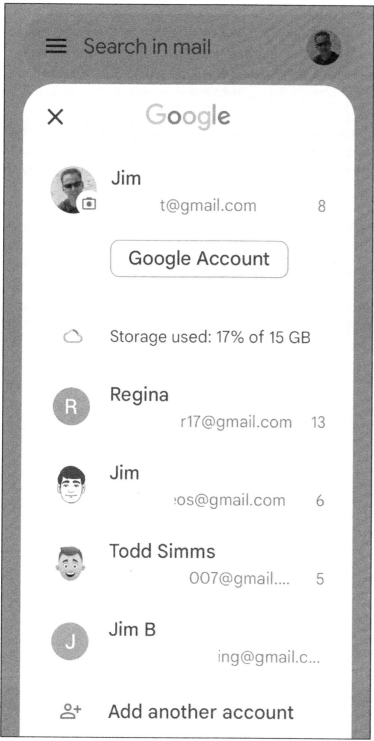

Figure 6.7

Using Vendor Specific Email Apps

The easiest way to set up your email is to use the app that the email vendor (provider) supplies that you can download and install from the Play Store. You can find these for most of the more popular email providers, such as Outlook (Microsoft), Yahoo, AOL, mail.com, iCloud, and more.

To use one of these apps, make your way to the Play Store and search for the email provider. Before downloading any apps, make sure that it is the official app for your email since many people will create their own apps that can be used with various email providers.

Figure 6.8 shows the Yahoo email app, and you can see that Yahoo provides this app since their name is shown underneath it as the company's name. Figures 6.9 and 6.10 show some of the other results that came up when searching for Yahoo email, and you can see that they are not provided by Yahoo itself, so this is the kind of thing you should be looking for. (Not that using another vendor's app is the wrong way to go about it, but you will most likely be better off using the app that is made by the vendor that provides your email account.)

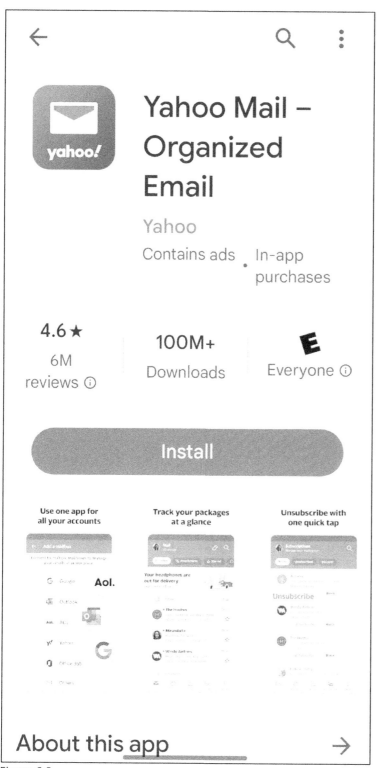

Figure 6.8

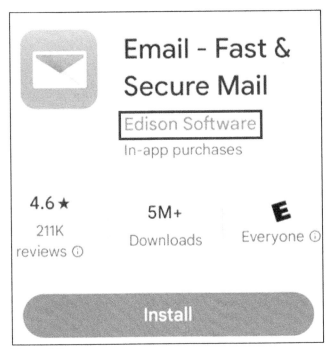

Figure 6.9

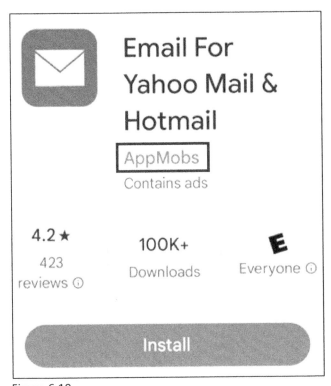

Figure 6.10

Once you get your email app installed, it's very easy to configure your account since all you need to do is enter in your username\email address and password, and it will configure the rest for you. These apps also tend to make the interface look similar to their website, so you won't end up needing to learn something new.

Chapter 7 – Browsing the Internet

Your smartphone allows you to connect to websites and do things such as read the news, check the weather, do some shopping, and so on, just like you can do on your computer at home or at the office. You can usually find an app specific to the task that you would like to perform online, such as checking the weather or shopping, but you really don't need to clutter up your phone with a bunch of single purpose apps if you just want to look something up online really quickly.

Android smartphones come with a built in web browser, and most of the time it will be the Google Chrome browser since it's owned by Google as well, so they want you to use their browser of course! This doesn't mean you can't use a different browser on your phone such as Firefox, Opera, or Edge, and just like with your computer, you can have multiple browsers installed at the same time.

Using the Google Chrome Built-in Web Browser
Since you most likely have Google Chrome installed, I will be using it for my examples in this chapter. Other browsers should work in a similar fashion, so it shouldn't be that hard to follow along if you are using one of them.

Figure 7.1 shows the Google Chrome browser connected to the google.com website. If you use the Google search engine at home, then you will notice the familiar headings such as News, Videos, Maps, and Images that can be used to search for those particular things.

There are a few parts of a web browser that you should be aware of, and they should be fairly common between different browsers. All browsers will have an address bar that shows the website address that you are currently on. If you know the address of a website you want to go to, then you can simply type it in this section and tap the enter key on your keyboard. What the enter key looks like will vary depending on your phone model and what keyboard you are using. Many times it's a checkmark, or even the word GO when using your web browser (figures 7.2 and 7.3).

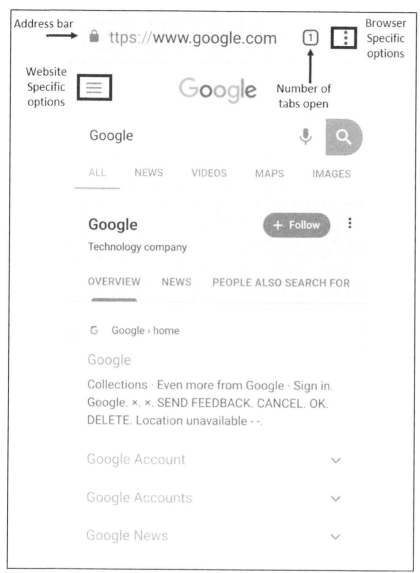

Figure 7.1

Figure 7.2

Figure 7.3

Looking back at figure 7.1, you will also see that there is a small box with the number 1 shown. This tells you how many tabs you have open within the browser. Tabbed browsing is where you can have multiple websites open within your browser at the same time, and each one is on its own tab. It works the same way as it does on your desktop computer, and if you aren't using it, you really should be!

To open a new tab in Chrome all you need to do is tap on the tab indicator (the number 1), and then click on *+ New tab* to open a new tab (figure 7.4). When you do this, Chrome will open a new tab and load whatever your home page is set to be. (I will be going over how to change your search provider and other settings later in this section.) For some reason, Google took away to option to change your home page, so your options are limited. In my case, my home page is set to Google, so from there I will search for *new cars,* and in the results open the website for Autotrader.com.

Figure 7.4

Figure 7.5 now shows the autotrader.com website open, and you will notice how the tab indicator shows a 2 because I have my original Google page open and now this Autotrader page.

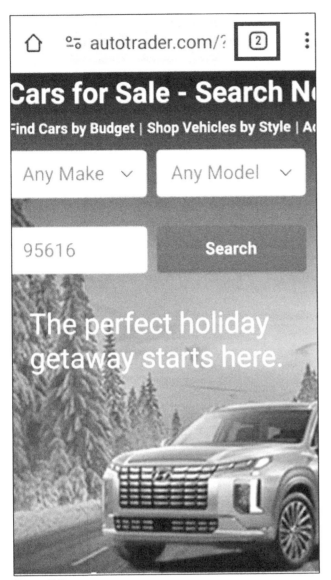

Figure 7.5

If I tap on the tab indicator again, it will show me both of the websites that I have open, allowing me to toggle back and forth between them (figure 7.6). You can do this process over and over and have many tabs open, but after a certain point, things tend to get a little messy and it gets difficult to stay organized.

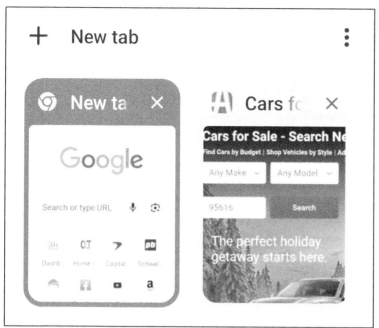

Figure 7.6

Going back to figure 7.1, you will see that there are three horizontal bars at the upper left of the web page. These are used for website specific options, and not all websites will have these. If you tap on them, you will be shown options similar to figure 7.7, which are for the google.com website. Depending on what site you are on, the choices from this type of menu will vary.

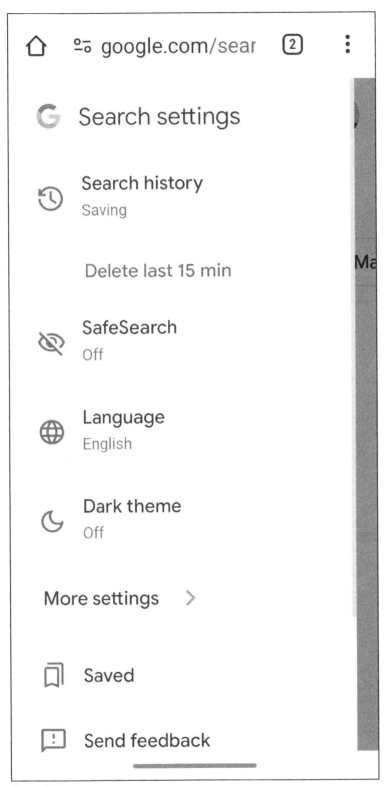

Figure 7.7

There are options specific to the Chrome web browser that you can access by tapping on the three vertical dots at the top right of the screen. Figure 7.8 shows the settings that you can configure for Chrome.

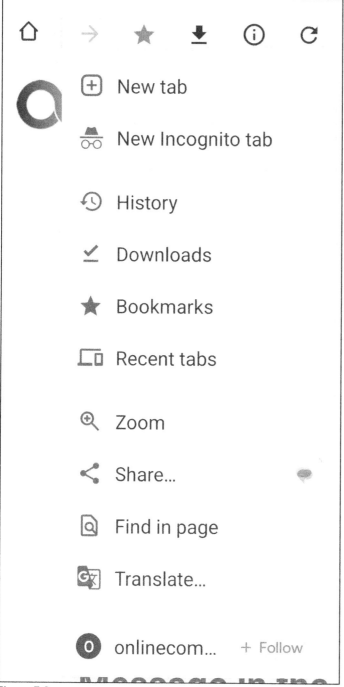

Figure 7.8

Many of the choices are self-explanatory, but I will quickly go over all of them so you have an idea of what each of them does.

- **New tab** – This will open a new tab just like tapping on the tab indicator button will do.

- **New incognito tab** – This will open a new tab in private browsing mode where Chrome won't save anything you do in its history.

- **History** – This will show you what websites you have gone to in the past, and it's also possible to visit those websites again right from your history.

- **Downloads** – If you have downloaded any files like pictures or documents from websites, you can view and open these downloaded files from here.

- **Bookmarks** – Here you can add bookmarks (favorites) for websites that you want to go back to at a later point in time. If you use Chrome on other devices and are logged in with your Google account, you can configure it to sync your bookmarks among any devices that you are logged into with your Google account.

- **Recent tabs** – This will show you a listing of tabs that you have had recently opened, but have now closed, making it easy to go back to one or more of them if desired.

- **Zoom** – This will allow you to zoom in and out of the page to make it larger or smaller.

- **Share** – When you come across a website that you want to share with others, you can choose the *Share* option and share that site via things such as email or text message, as well as copy the website address to paste into another app, or even print the page to a printer if you have one configured.

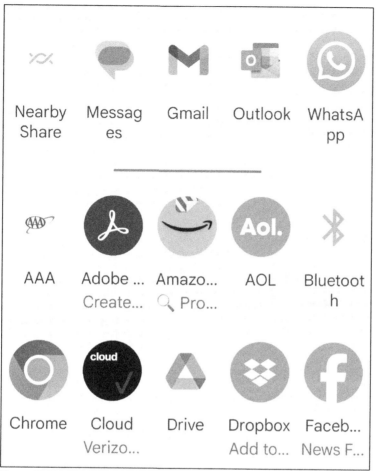

Figure 7.9

- **Find in page** – If you need to find a specific word or phrase on a webpage, you can search for it using this option.

Figure 7.10

- **Translate** – Many times you will go to a page that is in a different language and this will all you to have it translated to the language that is configured on your device.

- **Add to Home screen** – Using this option will add the specific website you are on as an icon on your home screen, giving you quick access to that website.

- **Desktop site** – Most websites have what they call a mobile version and desktop version of their websites, with the mobile version being formatted to look better on smartphones and tablets, and the desktop version formatted for your desktop computer. If for some reason, the mobile version is giving you issues or not displaying like it does for you on your computer, you can choose to have Chrome use the desktop version instead.

- **Settings** – Here you will find many of the same Chrome settings that you will see on your desktop computer. I won't go over all of the settings but want to mention a few of them that you should know about.

I mentioned earlier in this chapter how you can change your search provider if you don't want to use Google. All you need to do is tap on *Search engine* in the settings, and then choose another one from the options provided. Your choices might be a little different than mine.

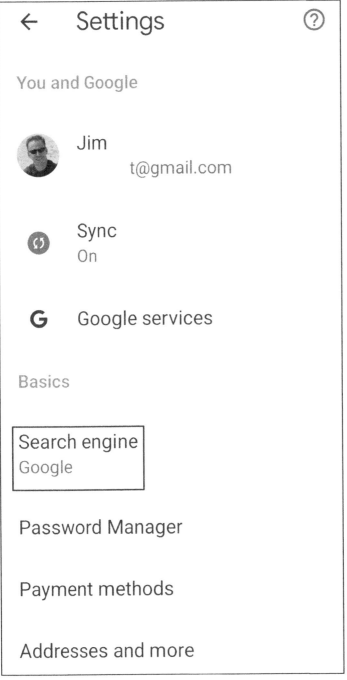

Figure 7.11

As you can see in figure 7.12, I have four other options to use as search engines. If you aren't a fan of Google, then you can try one of the others provided and always switch back if you are not a fan. Also, note that in figure 7.11 it says Sync is on which means it's syncing my browser information (such as bookmarks and history) to my Gmail account like I mentioned you can configure earlier if you log in with a Google account on all of your devices.

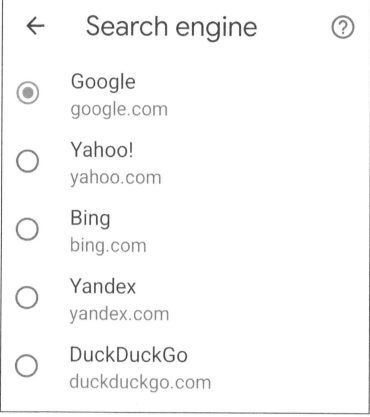

Figure 7.12

If you don't know what a search engine is, it's the site you use to perform searches on. Many people get search engines and web browsers confused, but the way to think about it is that you use a search engine via your web browser to search the Internet.

If you were to scroll further down the page on the main settings screen, you will see an option for *Privacy and security*. Here is where you can clear out your search history, website cookies, and images from Chrome, as well as set other options to keep your browsing a little more private.

Within the Privacy options, you will see a section called *Clear browsing data*. Here you can choose what you want to have removed and for what time period such as the last hour, last 24 hours, last 7 days, last 4 weeks, or all time. Just keep in mind that if you delete your cookies, you might lose some of your saved logins and have to re-enter them the next time you log onto that site. Then they will be stored in your browser again.

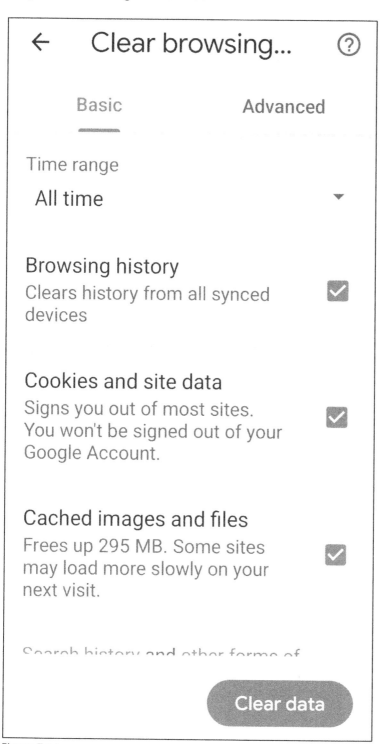

← Clear browsing... ⊘

Basic   Advanced

Time range

All time   ▼

Browsing history
Clears history from all synced   ☑
devices

Cookies and site data
Signs you out of most sites.
You won't be signed out of your ☑
Google Account.

Cached images and files
Frees up 295 MB. Some sites
may load more slowly on your   ☑
next visit.

Search history and other forms of

Clear data

Figure 7.13

Installing Other Web Browsers

Google Chrome is a great web browser, but it might not be the best browser for you. If you want to try out some others just to see if you can find something better for your needs, you can head back to the Play Store and download as many as you like. Or, should I say, as many as you have room for on your phone!

To install another web browser, you do the same process as you would for installing any other app. If you know the name of the browser you want to install, you can search for it by name. Otherwise, you can simply search for *web browser.* When you do this search, you will notice that you get a lot of results, and you may not know where to begin. This is where reading reviews and looking at the number of downloads will come in handy.

You might also want to go online and do a search for the top Android web browsers so you can get some recommendations from people in the industry who have tested these browsers and can vouch for them. Like I mentioned before, you can have multiple web browsers installed on your smartphone and use all of them at the same time.

If you find a browser you like better than Chrome, then you can go into your apps from the Android Settings and make it your default browser, which means that when you perform any action that opens a website (such as tapping on a link from an email) it will use the browser you have specified. This setting will most likely be under the *advanced* section.

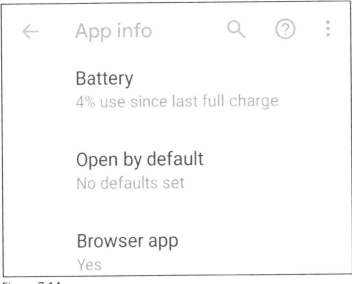
Figure 7.14

Bookmarks\Favorites

Just like with the web browser on your home computer, Android web browsers allow you to create bookmarks for sites that you want to go back to and visit at another time. A bookmark (also called favorite) is simply an entry for a website's address that can be easily referenced from a stored location within your browser.

When you are on a web page that you want to save so you can visit it later, all you need to do is have the browser create the bookmark based on that site's address and you are all set. Once again, I will be using the Google Chrome web browser to show you how to create and view your bookmarks, so the process will most likely vary in other browsers. Within the Google Chrome browser I am on a site called www.onlinecomputertips.com (which is actually my computer support site, so check it out if you are interested!). From there I will tap on the three vertical dots like I did to get to the Chrome settings, and from there I will tap on the star icon to add the page to my bookmarks.

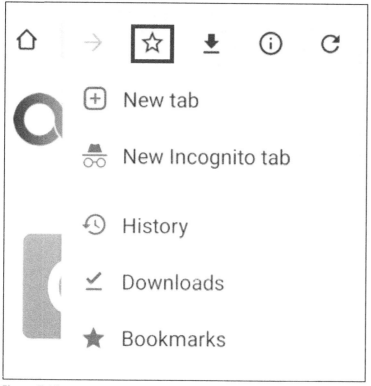

Figure 7.15

Chrome won't tell you that it added it to your bookmarks, but if you were to go back and tap on the three dots again, you will notice that the star is now filled in, indicating that the page is in your bookmarks.

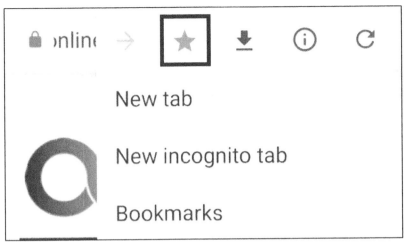

Figure 7.16

Now if I go to my bookmarks (which is done by tapping on the word *Bookmarks* as shown in figure 7.16), I will see what websites I have created bookmarks for. As you can see in figure 7.17, I have *Mobile bookmarks* with one bookmark, and *Bookmarks bar* with 151 bookmarks. This is because I have my Google account synced with my phone, so it's taking all of my bookmarks from my desktop computer and synchronizing them with my phone.

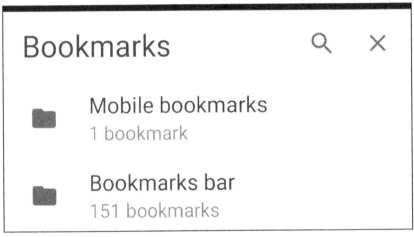

Figure 7.17

If I were to open Mobile bookmarks, I will see the new bookmark I created for my computer help site (figure 7.18). As you can see, I tend to create all of my bookmarks on my desktop PC and don't really go anywhere new on my phone that I care to save for later.

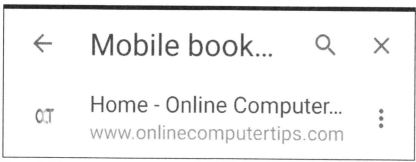

Figure 7.18

Sharing Websites with Others

I had mentioned earlier in this chapter that you can share websites with other people using various methods but wanted to go over a couple of examples of how to do this in case this is something you think you might want to do, and it most likely will be!

The easiest way to share a website is to use the share option from within your web browser. This can usually be found from the menu settings, or sometimes you will see an icon that looks like the image below, which is the universal symbol for sharing.

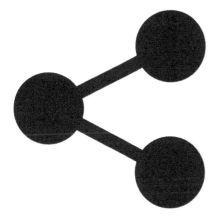

Once you find the sharing option within your browser, you can choose how you want to share the link. As you can see in figure 7.19, there are many ways to share a website link, and the options you will have will depend on what apps you have installed on your phone.

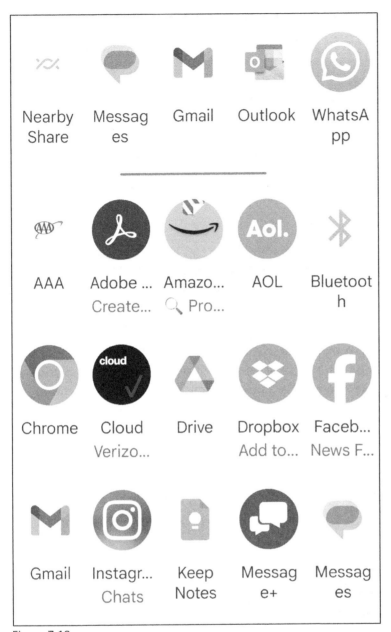

Figure 7.19

For my first example, I am going to share a website via text message, so I will choose the Messages icon from the list. Then I will pick an existing conversation to add the link to or create a new one. I can send it to one person or several people at the same time (I'll be going over that in Chapter 12). Figure 7.20 shows the results.

Figure 7.20

For my next example, I will send the link via email and use my Gmail account to send it off to someone. This time I will choose Gmail from the list of available apps to share with and enter the email address of the person I want to send the email to. It will add the website name as the subject. Then all you need to do is simply send it off and that's it!

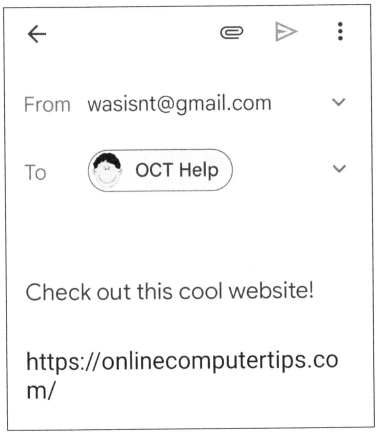

Figure 7.21

Then the person on the other end would get the email with a link that they can click on.

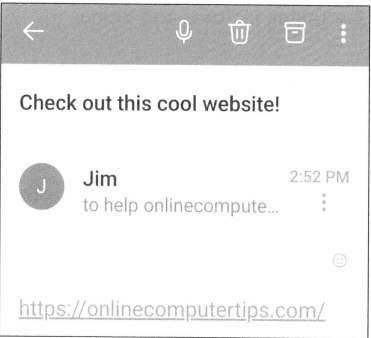

Figure 7.22

Chapter 8 – Taking Pictures and Videos

Taking high quality pictures and videos is one of the best parts about having a smartphone, and the nicer the phone, the nicer the camera will usually be. Thanks to smartphones, there is no real need to own a digital camera anymore unless you are a professional photographer who needs capabilities above and beyond that of a phone camera.

Android smartphones will come with a built-in camera that is capable of taking pictures as well as recording videos. For the most part, you will have a camera on the front of the phone for things like selfies, and one on the rear for taking pictures of anything you find to be picture worthy. These pictures and videos will be stored on your phone, and from there you can do things like share them with others, post them to social media sites, and even copy them to your computer.

Using the Camera
For the most part, it's very easy to use the camera that came with your phone. All you need to do is find the icon that opens the camera itself, point it at your subject, and then press the shutter button. Many cameras will even have shortcuts to open the camera such as pressing the screen on button twice in a row quickly. For the most part, you can find your camera icon at the bottom of your main home screen.

Once you open the camera app, you can choose from taking photographs or videos, and you will have other options that will vary depending on what phone you have. (Your phone should default to photos rather than videos when you first open the camera.) If you take a look at figure 8.1, you will see the various parts of the camera interface. Yours will most likely vary a bit if you have a different phone.

The main thing you should be concerned with is where the shutter button is that actually makes your phone take the picture. It should be right at the bottom middle section of the screen. By default, your phone should use the rear camera to take pictures of things in front of you, but if you want to swap cameras to the front facing one, then you should have an option similar to what is shown in figure 8.1.

Speaking of options, you will most likely have some options at the top of the screen, which I will go over shortly. One option you will use all the time is the zoom, and you can either use the zoom button if it has one, or you can pinch to zoom right on the screen itself.

Figure 8.1

Another important thing to know is that once you take a picture, you should have an option to view the last picture that was taken right from the camera app itself, which makes it easy to see if the picture is one you want to keep, or one that you think should be retaken.

When it comes to recording a video, all you need to do is switch the camera mode from photo to video, aim at what you want to record, and press the record button.

You will see the elapsed recording time as you are filming (figure 8.3). This is also a good way to know that you pressed the record button for sure! Then when you are done, simply press the button again to stop the recording.

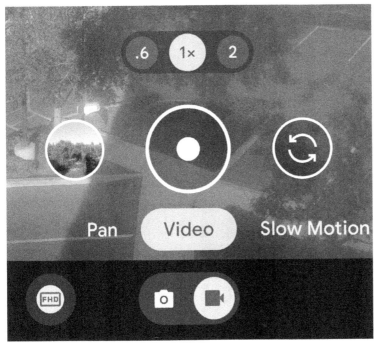

Figure 8.2

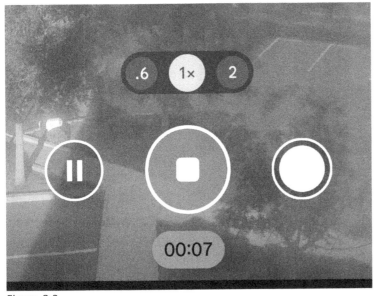

Figure 8.3

Up at the top of your camera, you may have some options to do things like adjust the light balance or settings for what type of environment you are using your camera in. You might also have a settings icon that you can tap on to get to these adjustments as well. Figure 8.4 shows some of the choices I have when I go to the basic settings. I can tap on *More settings* to get additional options.

Figure 8.4

All smartphone cameras should have an option to set a timer on the shutter so that you can have a delay for when the picture will be taken. This comes in handy if you are setting up a shot that you want to be in and need time to make it from where your phone is set up for the picture and where you need to be in the picture. Figure 8.5 shows the options I have for the timer, and they include a 3 second and 10 second delay.

Figure 8.5

Once you set the delay time, all you need to do is set up your phone where you want it to be for the shot and press the shutter button. Then it will do a countdown and automatically take the picture when the countdown is complete. You should either have a visual display on your screen like figure 8.6 shows or have some sort of beeping sound indicating the countdown interval before the phone takes the picture.

Figure 8.6

Smartphone cameras will also come with a built-in flash that you can either set to *on*, which will use it for every picture, *off*, which won't use it at all, and *auto*, which will use the flash when needed. The flash settings can usually be found at the top of the camera screen with the other settings.

Figure 8.7

Many of the higher end cameras will come with additional features and effects to enhance your pictures or add fun elements to them. Figure 8.8 shows some options that some cameras have, such as creating slow motion videos and a photo booth effect.

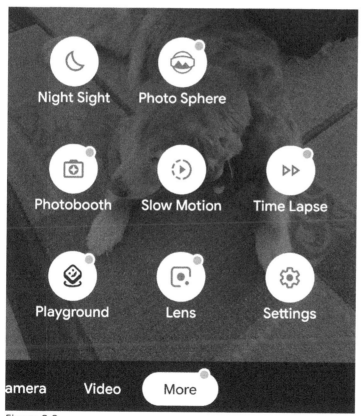

Figure 8.8

Camera Settings
You might be getting sick of reading about settings by now, but be warned that your camera will come with its own settings that allow you to fine tune how it works. Of course, you can just leave the settings at their defaults and probably be fine, but if you are the type who likes to "fiddle" with things, then you might want to check them out.

Once you get into your settings, you can see what you have to work with. Once again, your settings will probably vary from mine, but probably not by much. Figure 8.9 shows some of the general settings that I have for my camera.

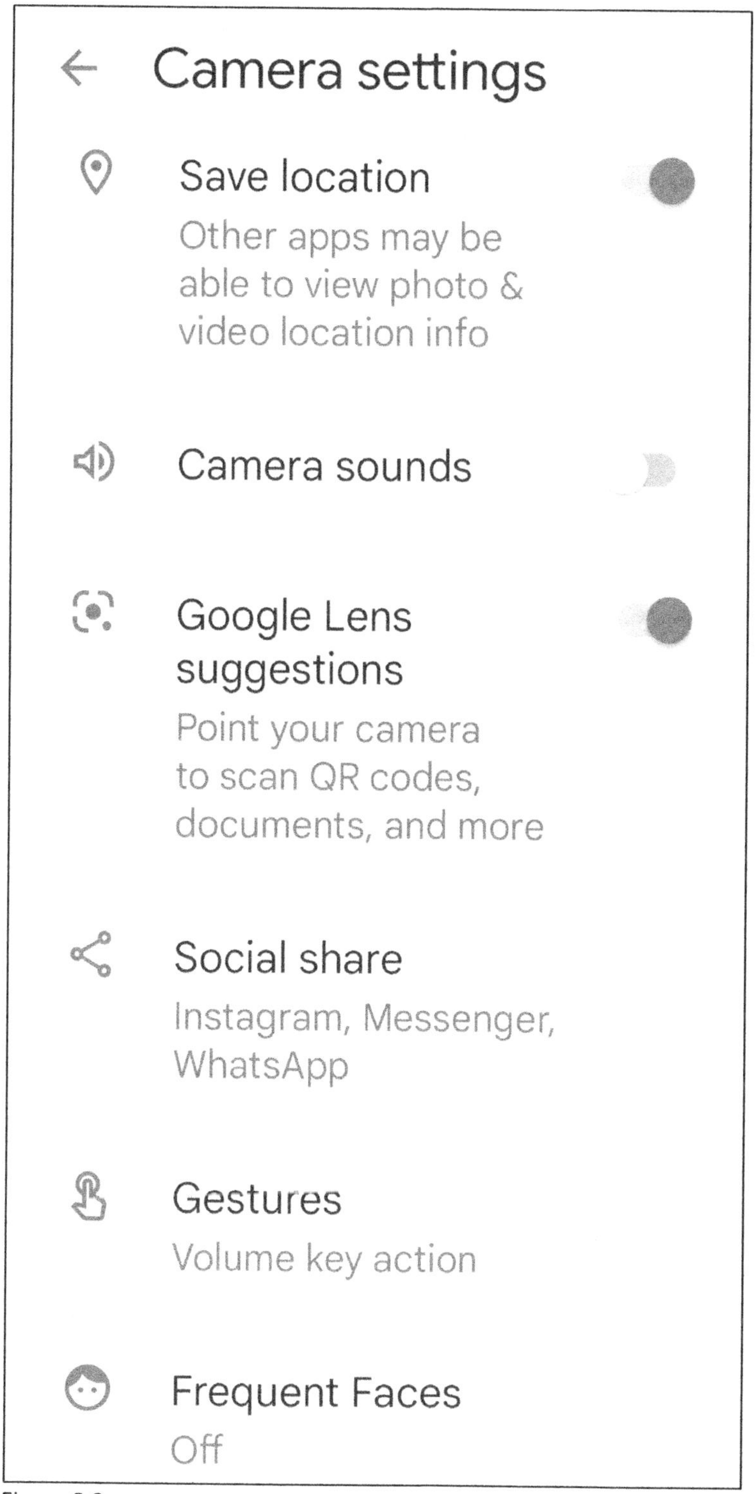

← **Camera settings**

Save location
Other apps may be
able to view photo &
video location info

Camera sounds

**Google Lens
suggestions**
Point your camera
to scan QR codes,
documents, and more

Social share
Instagram, Messenger,
WhatsApp

Gestures
Volume key action

Frequent Faces
Off

Figure 8.9

The *Save location* setting is used to add what is called a "tag" within your picture to keep a record of where a picture was taken. This comes in handy later on because many apps and computer programs can read these tags, which allows you to see where you were when a certain picture was taken in case you forgot.

When you take a picture, your phone will usually make a camera shutter sound, and if that is something you don't want to hear, then you can turn it off from the settings.

Google Lens is a Google specific feature that allows your camera to detect what it sees and then provide you with more information on that object. So, if you take a picture of some text, it will allow you to edit that text and use it in another application. Or if you take a picture of a certain product, you can use Google Lens to find that product online so you can compare prices. Google Lens comes standard on Google Pixel smartphones and some others but can also be downloaded from the Play Store.

Gestures are used to make common camera actions faster and more convenient. For example, you can set your volume key to be the shutter button to take the picture, or double tap on the screen to quickly zoom into the thing you are taking a picture of.

Many cameras will place a virtual grid over the camera screen which helps you better align your pictures if you are the perfectionist type. This grid can be turned on or off as needed.

The resolution of your camera will depend on what make and model you buy, with the more expensive ones having better cameras, of course. You don't need to stick with the default resolution, and you can change it as needed. The higher the resolution, the more storage space a picture will take on your camera. You can adjust these settings as well as the aspect ratio of your pictures for the front and rear cameras from the settings area. If your camera has built-in video stabilization, then you can turn that on and off from here as well.

Audio zoom is used to amplify the recorded sound to make it louder when zooming in while taking a video. You can think of it as zooming in your microphone. I find that it tends to degrade the quality of the audio when using this feature.

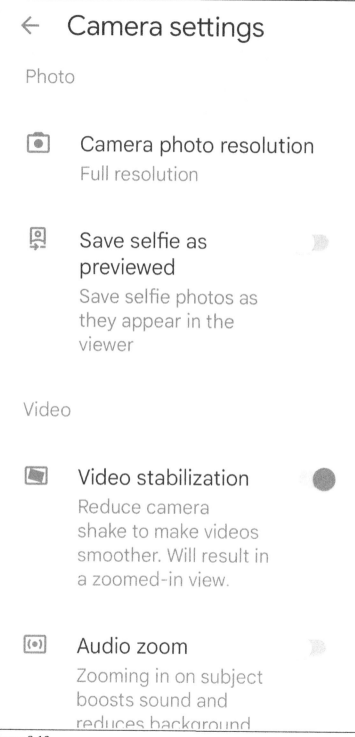

Figure 8.10

When changing resolution and aspect size, it's always a good idea to test your new settings before assuming they will be what you are looking for, otherwise you might end up with a bunch of pictures from your vacation not looking the way you had expected!

Viewing and Organizing Your Pictures

Now that you have taken a bunch of pictures from various events, you might want to organize them so they make a little more sense. How you do this will depend on what photo app you are using.

Most Android Smartphones will have the Google Photos app built-in, which works fine for keeping your pictures organized, but there are better apps out there, so if you are not happy with Photos, then feel free to try out some others. Since you should already have the Photos app, that is what I will use for my examples in this section.

Figure 8.11 shows what the Photos app looks like when you open it. It will group your pictures and videos in a thumbnail by the date taken, and also add their location if that feature is enabled on your phone (which it should be). Notice on the bottom of figure 8.11 that there are sections for Photos, Search, and Library.

To view a photo all you need to do is tap on it to make it full screen. Then you can swipe left to see the next one, or swipe right to see the previous one. To get back to the thumbnail view, you can tap the back button on your phone.

If you want to delete a picture or pictures simply long hold on a picture until it puts a checkmark next to it. Then you will notice at the bottom of the screen you will have a trashcan option that you can press to delete the picture(s) (figure 8.12). Notice how you have other options such as sharing or the + sign, which can be used to add those pictures to an album. If you change your mind, you can tap on the checkmark to uncheck a picture or pictures.

Figure 8.11

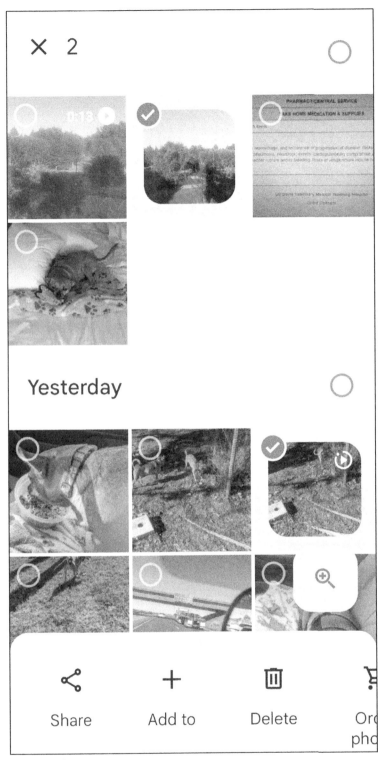

Figure 8.12

Tapping on the *Library* button at the bottom will show you what albums you have on your phone. You will notice that Photos likes to organize things for you and even create groups on its own based on what it sees in your pictures. You might find that Photos creates groups based on what it sees in your pictures such as *People & Pets* or *Places*. It also shows the categories of photos on the phone such as general camera photos, screenshots, and pictures you have downloaded.

At the very bottom of the screen, you will see the actual albums that are on this phone. Figure 8.13 shows the albums I have in more detail, and I can tap on any of them to view their contents.

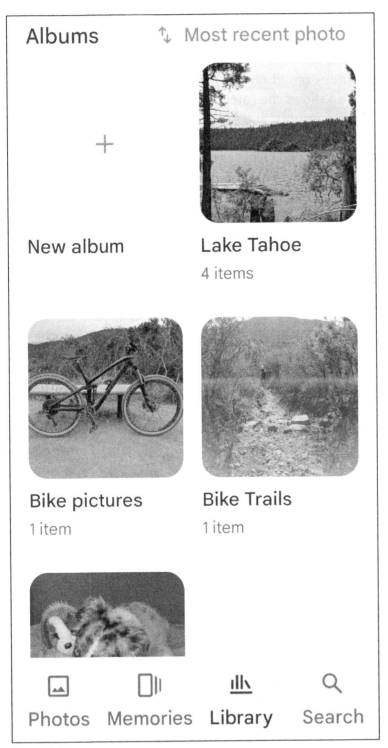

Figure 8.13

Figure 8.14 shows how an album looks when you open one. You will then have additional options such as sharing your album or adding additional photos to it.

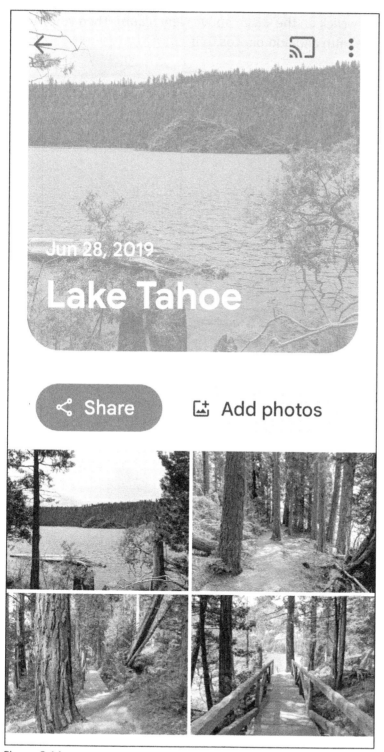

Figure 8.14

To add a new album simply click on the + sign above *New album.* Then you will be asked to name your new album and add photos to it.

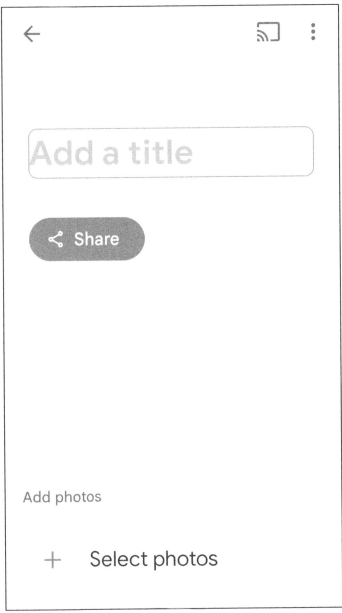

Figure 8.15

For my example I am just going to create a regular album, call it *Dog Pictures*, and add three pictures from my existing photos. Figure 8.16 shows the process for adding the three pictures, and figure 8.17 shows those four pictures with the *Dog Pictures* title I gave to the album.

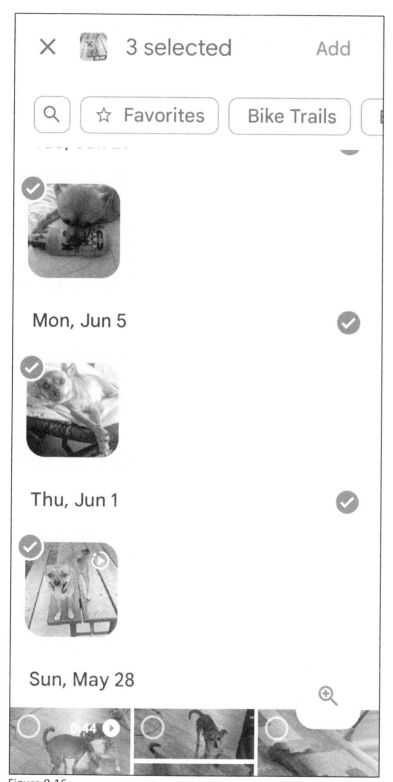

Figure 8.16

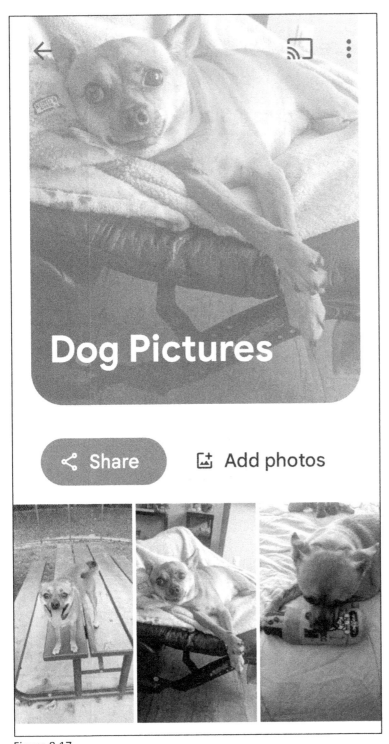

Figure 8.17

Finally, figure 8.18 shows my new album listed with my existing albums.

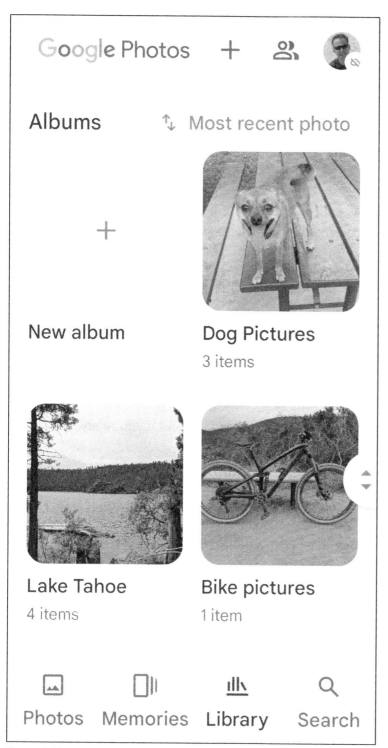

Figure 8.18

Be aware that Google likes to keep your photos and videos in your account even if you delete them from your phone, so if you use a different app than the Photos app and remove pictures, they might still remain within your Photos account. You can check this by comparing your other app with the Photos app.

Checking Your Phone's Free Space

As you take more and more pictures (and videos), you might notice that your phone will start to run low on space because these high resolution photos and videos are larger than they were back in the day of the subpar cameras that came with older phones.

Once you run out of space on your phone's storage, then you won't be able to take any more pictures or even do things like install new apps because your phone won't have anywhere to put them. The easy fix is to either delete pictures you don't want any more or transfer them somewhere else like to your computer and then remove them from your phone to free up space. Another option is to uninstall any apps that you don't need or can live without if you really don't want to remove any pictures from your phone.

To check the available space on your phone, you will need to go back to the Settings app and look for a section called *Storage*. From there, you will be shown your total storage and how much space you have used and free. Depending on your phone, you might also be shown what types of items are using up what percentage of your storage.

You can use this information to help determine what the best method for freeing up space might be. If you have an option similar to the *Manage storage* button that I have shown in figure 8.19, then that can be used to remove unused apps, temporary files and downloaded files that you might not need anymore. (Just be sure you know what you are deleting before using this option.)

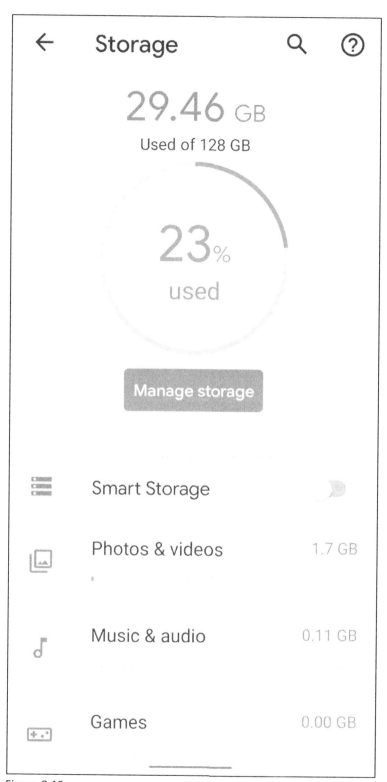

← Storage 🔍 ⑦

29.46 GB

Used of 128 GB

23%

used

Manage storage

Smart Storage

Photos & videos   1.7 GB

Music & audio   0.11 GB

Games   0.00 GB

Figure 8.19

Transferring Pictures and Videos to Your Computer

If and when you do decide that you want to transfer pictures and videos from your phone to your computer, it's a pretty easy process to do. Just keep in mind that you can either COPY pictures from your phone to your computer, or you can MOVE them. Copying them won't free up any space from your phone but moving them will.

The first step in the process is to connect the USB cable that came with your phone to the normal charging port on your phone, and then to a free USB port on your computer.

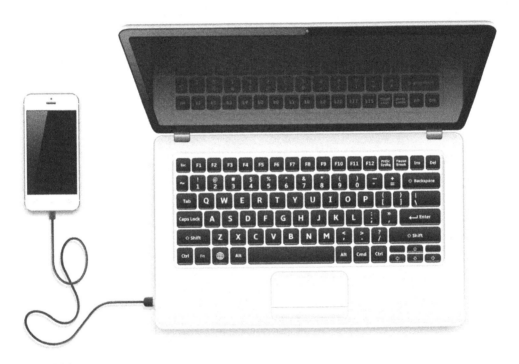

Figure 8.20

The next step involves telling your phone that you want to use the connection to your computer to transfer files. This is usually done by pulling down from the notification area, tapping on the USB section (figure 8.21) to open up the connection options, and choosing the appropriate action (figure 8.22).

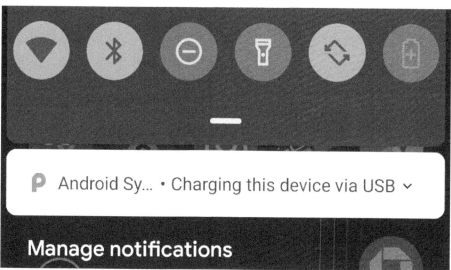

Figure 8.21

Notice in figure 8.22 that I chose the *File Transfer* option because I want to transfer files from my phone to my computer. You may see options with slightly different names such as *photo transfer,* for example.

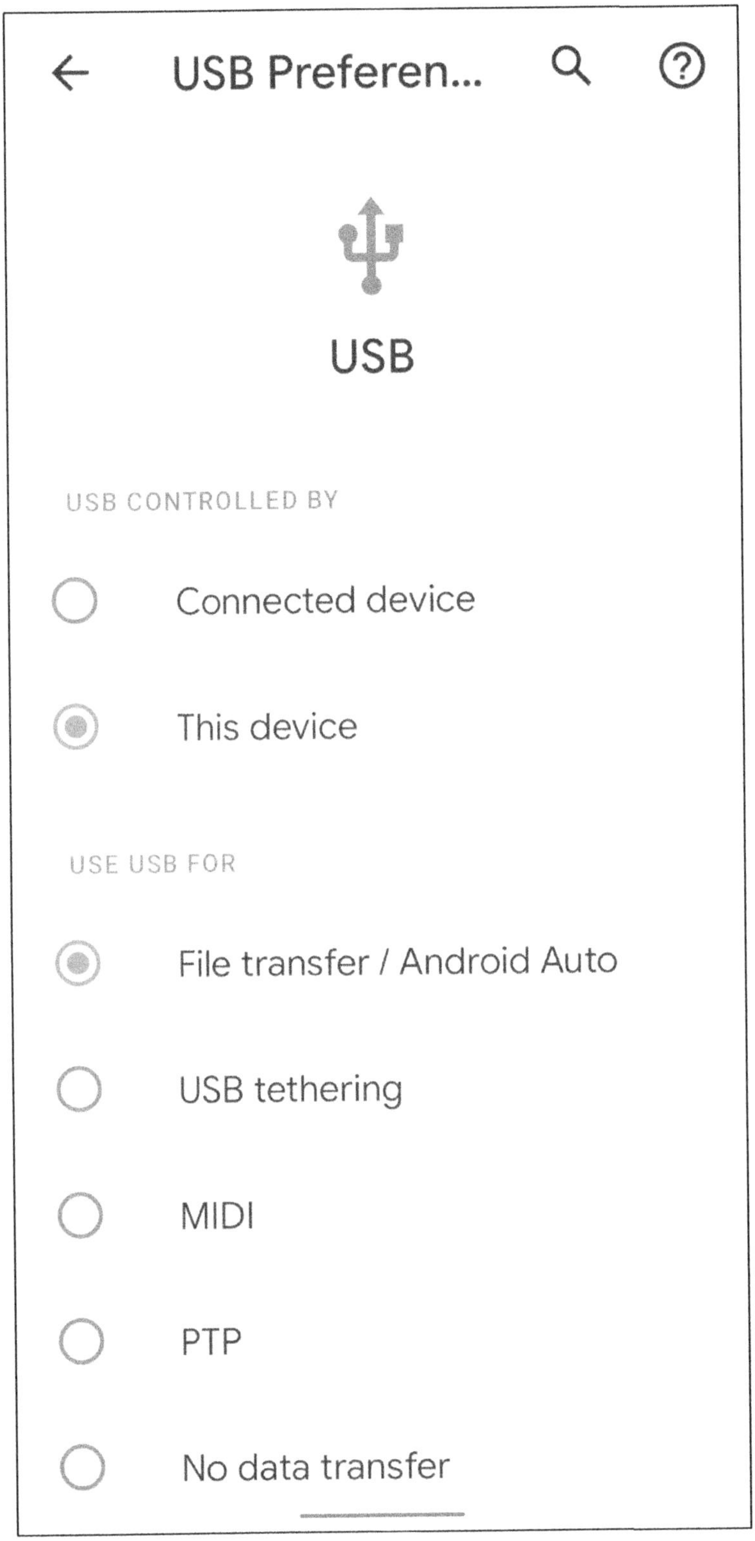

←   USB Preferen...   Q   ⑦

USB

USB CONTROLLED BY

○ Connected device

◉ This device

USE USB FOR

◉ File transfer / Android Auto

○ USB tethering

○ MIDI

○ PTP

○ No data transfer

Figure 8.22

Then on my computer (I'm running Windows), I should see my phone appear. Then I can double click on its internal storage to see the files and folders contained on my phone.

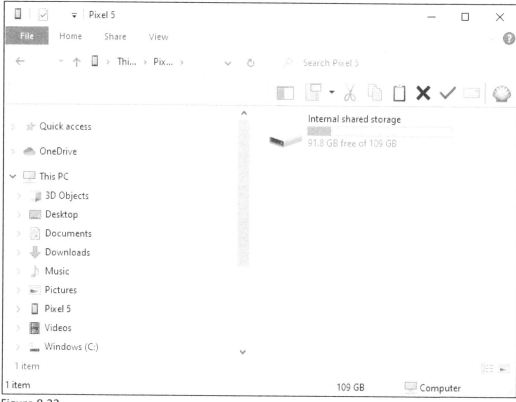

Figure 8.23

The folder I want to look for is named DCIM, and when I find that, I want to double click it to open it up.

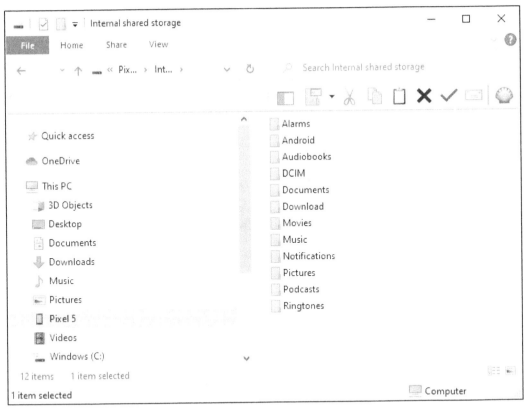

Figure 8.24

Within that folder you may see your pictures, or you might have another folder called *Camera* that you will need to open up. Once you are here, you can drag and drop the pictures from your phone to your computer and then delete them off of your phone after you confirm that they have been copied over. You can delete them using the phone, or you can delete them right from this DCIM folder that you opened up on your computer.

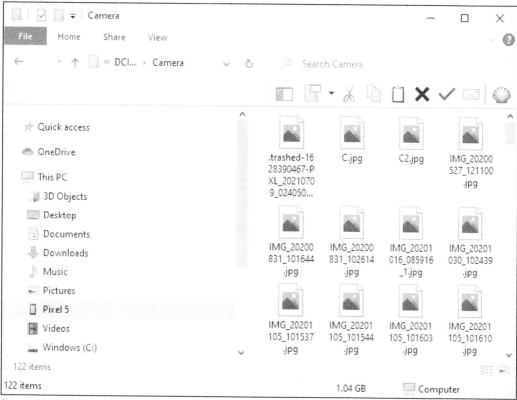

Figure 8.25

If you are looking for pictures that were sent via text message or saved from taking screenshots, then you can look for a folder named Pictures rather than DCIM and inside that folder you should find what you are looking for.

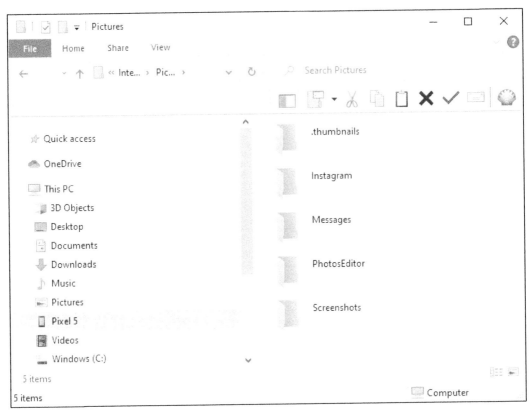

Figure 8.26

Chapter 9 - Listening to Music

Another one of the more common things people use their smartphones for is to listen to their favorite music. If you are old enough to remember Walkman's, then you know how long people have been using portable music devices so they can have their favorite tunes playing everywhere they go.

Many people still use MP3 players such as the Apple iPod and other devices to listen to their music on the go, but these are slowly (or maybe quickly) being replaced by smartphones since you can use your phone as your MP3 player and still have all the functionality of your phone without having to carry around two devices.

Adding Music to Your Phone
If you have MP3 or other audio files on your home computer, then you can copy them over to your phone so that you can listen to them anywhere you like. This process is similar to how you copy pictures *off* your phone, but instead of moving files *from* your phone, you are adding them *to* your phone.

To accomplish this, you need to connect your phone to your computer just like you did in the last chapter for the section called *Transferring Pictures and Videos to Your Computer*. But instead of looking for the DCIM folder, you will look for a folder called *Music* (figure 9.1). Then you will drag your MP3 files (or folders) to this Music folder to copy them over.

Keep in mind that music files take up space just like photos and videos, so you will need to check your storage space if you think you might be copying over too many music files to your phone.

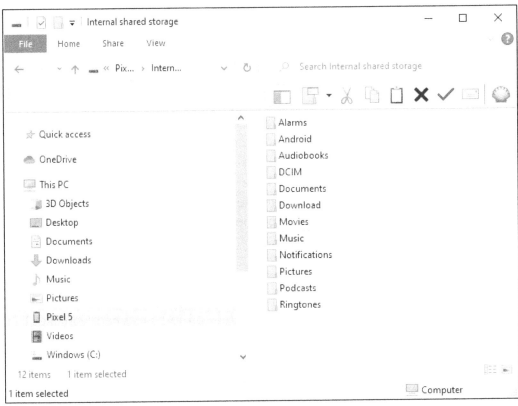

Figure 9.1

Once you have your MP3s copied over, you can open up your music player app and it should find the files and allow you to play them.

YouTube Music App

Your phone may or may not come with a third party music player app installed, but it should come with the YouTube Music app installed since it's made by Google for Android devices.

When you open the YouTube Music app, it will ask you to sign in for the first time if you wish to access any music you have stored online. Otherwise, you can choose the *device files only* option.

Figure 9.2

If you choose to sign in with your Google account, you will then be able to check out new music (figure 9.3) and also listen to your uploaded music if you have any (figure 9.4 and 9.5).

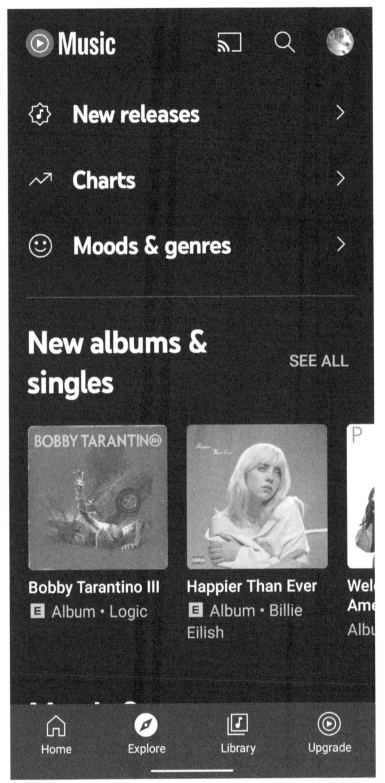

Figure 9.3

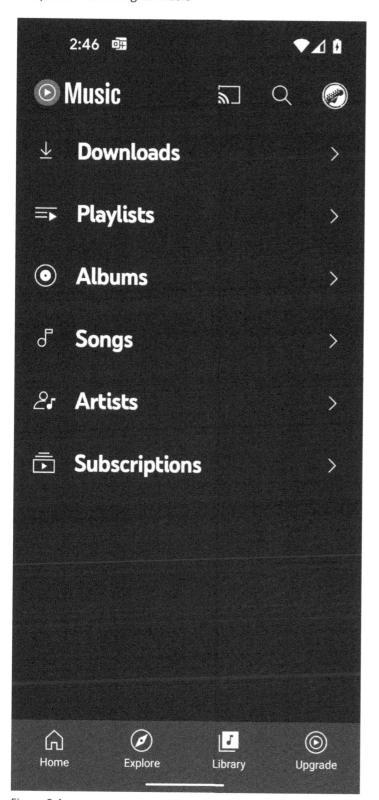

Figure 9.4

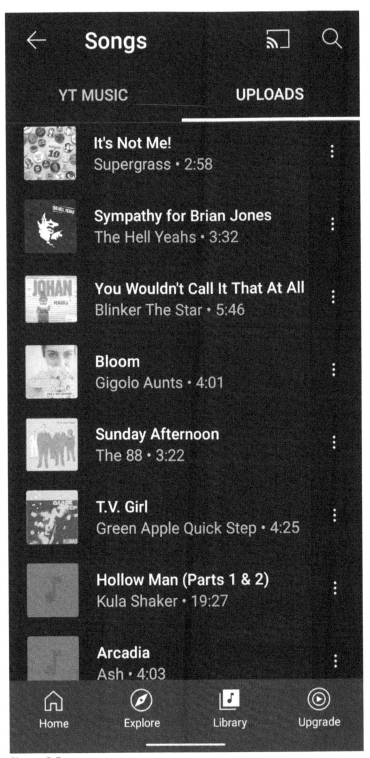

Figure 9.5

Streaming Music
YouTube Music will allow you to upload the music from your computer to your Google account so you can stream it from anywhere you have an Internet connection. Setting this up is beyond the scope of this book, but you can go to **https://music.youtube.com/** and get it configured pretty easily.

Once you have your music uploaded to YouTube Music, then it will show up in the categories seen in figure 9.4 alongside any music that you copied over to your phone directly. The main difference will be that it doesn't take up space on your phone because it's stored on Google's servers online and not directly on your phone. This way you can store way more music online than you can on your phone and not have to worry about running out of room.

One thing you do need to realize when streaming music is that it will use up data on your cellular phone plan if you are not connected to a wireless connection. So, if you don't have unlimited data, you might want to think twice about doing this. (I will be going over how to check your data usage in Chapter 16.)

There are other options to stream music on your phone besides having to use your own music or go out and buy the music you want to hear. You can use a streaming service such as *Pandora* to create stations based on artists you like. You won't be able to choose exactly what music you want to listen to though, because it will choose it for you based on the channel you create.

If you are looking for unlimited music streaming and are willing to pay for it, then you can try something like Spotify, which is a subscription-based streaming app. They do offer a free plan, but it will only let you stream 20 hours of music a month. If you want unlimited streaming, it will cost you about $10/month. Plus, you can use your account on multiple devices such as your smartphone and home computer.

Of course, there are other streaming services out there, so you just need to look around and find what works best for you in your price range. Regardless of the service you use, just remember that it will be using your data if you are not connected to a wireless connection like you would be at home or at the office.

Chapter 10 – Contacts

Since you use your phone to do things such as make phone calls, send emails, and text other people, it makes sense to store contacts for all these people on your phone so you don't need to memorize them or carry around a separate address book.

Since you are logged into your phone with your Google account, all of your contacts will be stored with your account, making them accessible from anywhere you can log in to your Google account with. So, if you lose your phone or get a new phone, you don't need to worry about having to add all of your contacts again. And if you are on your home computer, all you need to do is go to the Google Contacts website and you can access everything from there.

To get to your contacts all you need to do is find the Contacts app icon, which will most likely be on the bottom of your home screen. When you open it up, you will see your contacts as shown in figure 10.1 (unless you don't have any yet, of course).

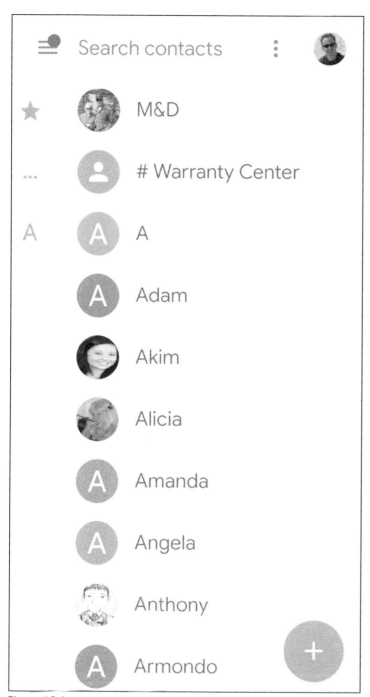

Figure 10.1

Adding Contacts

To add a new contact, all you need to do is tap on the + button shown in figure 10.1. Then you will be brought to a screen where you can enter the new contact's information. As you can see in figure 10.2, there are many fields you can fill in to enter a wide variety of contact information. You do not have to fill in all of them to create a contact. When you are done filling in the information, simply tap on *Save*.

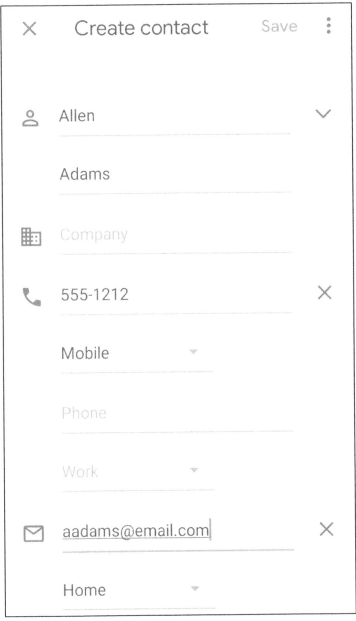

Figure 10.2

Your newly created contact will now show up in your list of contacts.

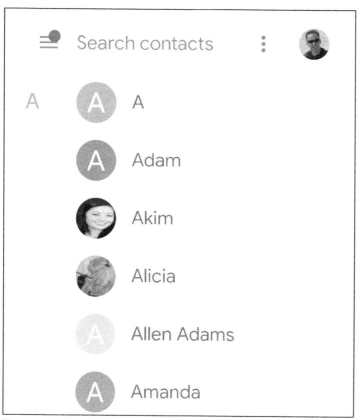

Figure 10.3

To open a contact, simply tap on it and you will see the information you have entered for that contact, plus actions you can take on that contact such as call or text. You can also edit the contact to make any changes by tapping on the *Edit contact* button.

Tapping on the star icon will make this contact a favorite, and it will then show up at the top of your list. Tapping on the three dots will give you additional options for that contact such as deleting it, sharing it, giving it a custom ringtone, blocking the number, and so on.

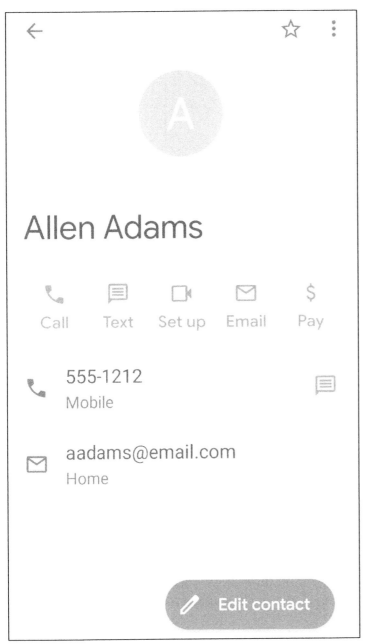

Figure 10.4

Using Contacts

If you want to call, text, or email a client (etc.), you can do so right from the contact (like shown in figure 10.4). But you don't need to go into your contacts each time you want to perform an action with one. For example, if you are in your Gmail app and want to email someone from your contacts, you don't need to exit out of

Gmail, go to your contacts, find the contact, and then tap on email. You can simply access your contacts from the Gmail app itself. Figure 10.5 shows an example of how you would do that. After clicking on *Compose* to start writing a new email, I simply tapped on the three vertical dots at the top right corner of the new email and chose *Add from Contacts*. This will then open my contacts, allowing me to choose anyone I like from my list. This will work on replying to and forwarding emails as well.

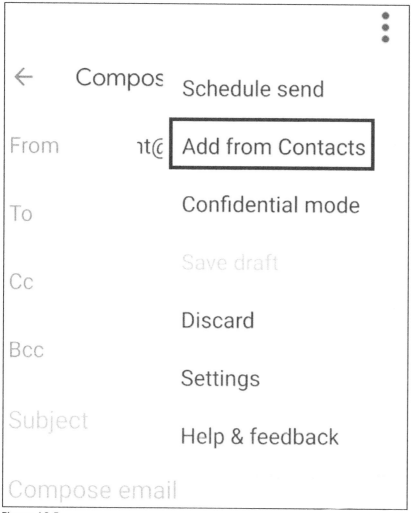

Figure 10.5

You can do this process from any app that uses your contacts to send things such as email, text messages, pictures, and so on. The way you access your contacts will vary between apps, so keep that in mind.

Adding Pictures to Your Contacts
Android smartphones give you the ability to customize your contacts by adding a picture to their contact entry if you desire. Adding a picture to a contact is a very easy process which I will now go over.

I recommend getting the picture you want to use for the contact on your phone first since you will need to specify where it is when you assign it to the contact.

The first step is to go to the contact you want to add the picture to and then edit that contact. From there, you should see an icon that looks like a camera (as shown in figure 10.6). This is what you will tap on to start the process of adding a picture to your contact.

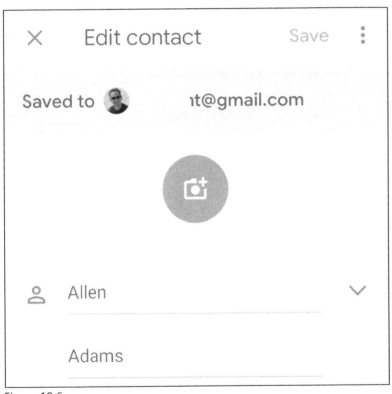

Figure 10.6

Next, you will be asked if want to take a new photo to be used for the contact or choose from an existing photo on your phone. You will most likely pick the *choose photo* option unless that person just happens to be there with you allowing you to take their picture on the spot.

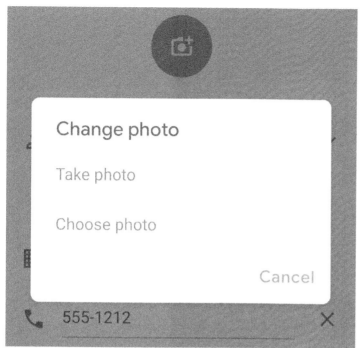

Figure 10.7

Once you choose the photo from your existing pictures (or take a photo on the spot) you will be prompted to edit the picture to make it fit better on the contact page. Simply drag the box with your finger to resize it, position the picture in the center of the box, and then tap on *Done* (figure 10.8).

Figure 10.8

As you can see in figure 10.9, I have the new contact photo setup for the Allen Adams contact. Then when I go back to my main contacts list, you will see that contact photo there as well (figure 10.10).

Figure 10.9

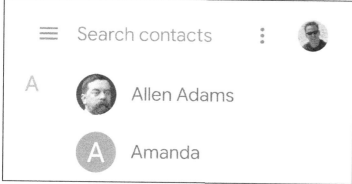

Figure 10.10

Chapter 11 – Making Phone Calls

Believe it or not, people actually use their smartphones to make phone calls, even though you wouldn't think so with how you are always seeing people staring blankly at them at all times! Making a phone call with your smartphone is a pretty easy thing to do, and after you do it the first time, you should be a pro!

Back when we had our flip phones, we would have the phone keypad right there when we opened our phone up. Smartphones work the same way, except we just need to take one more step before we see the number pad on the screen. Plus, there are many other improvements that smartphones have over flip phones when it comes to making phone calls.

Using the Phone Dialer Keypad

Obviously, you need to be able to dial a number if you are going to make a phone call, so this concept should be nothing new to you. So, the first step you need to take to dial a number is to have somewhere to do it from! You can usually find this number pad by tapping on the phone icon from your home screen.

Depending on your phone, when you tap on the phone app it might bring you to your frequent calls, favorites, recent calls, contacts, or some other section, but either way, you should be able to find the icon for the number pad.

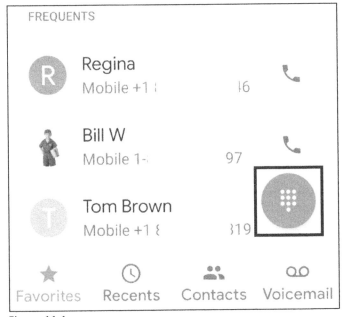

Figure 11.1

Once you open the number pad, you will see the standard numbers you can use to dial from and also the # and * keys. To make a phone call, simply dial the phone number and press the green talk button

Figure 11.2

Once the call is in progress, the green talk button will turn to red, and to hang up the call, simply tap on the red button. Figure 11.3 shows a call in progress. At the top of the screen, it shows the name of the person who is being called because that person is in my contacts. It will also show the image that I assigned to that contact.

Under the name is the call duration, which is how long the call has been going on. In this example, it's only been three seconds. Under that, we have several other icons to choose from. The *Mute* button will mute the call so you can hear them but they can't hear you. You will need to press it again to unmute the call. The *Keypad* button will bring up the number pad again in case you are making a call where you need to press a number for an automated menu. Pressing *Speaker* will put the call on speakerphone mode so you don't need to hold it up to your ear.

Add call is used to add another caller to the conversation, making it a three way call. *Video* call can be used to make video calls where you can see the person on the other end during the call. (The phones on both ends of the conversation need to support this feature for it to work.)

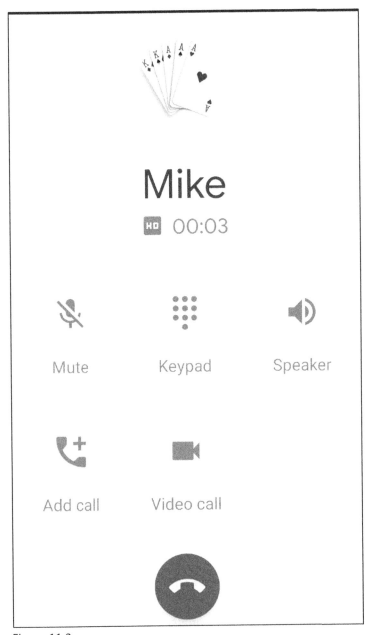

Figure 11.3

Calling a Contact

If you want to call someone from your contacts, then you don't need to actually dial their number, otherwise what is the point of having contacts! I went over contacts in the last chapter but will go over it again as a refresher here. From your phone app, you should be able to access your contacts, and they should look the same as they do if you access them from your contacts app.

Take a look back at figure 11.1 and notice how at the bottom of the screen there is a Contacts section. If you tap on this, you will be taken right to your contacts, just like you saw in Chapter 10. From there, all you need to do is tap on the person you want to call and then tap on the Call icon to start the call (figure 11.5).

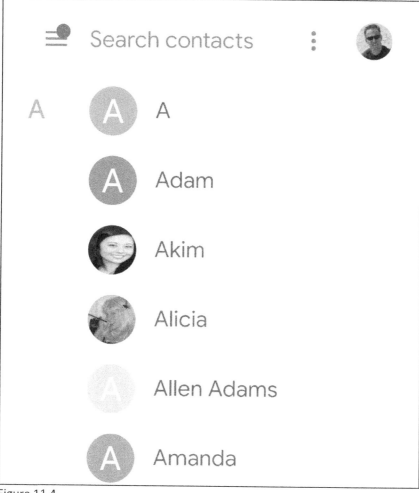

Figure 11.4

Figure 11.5

Checking Voicemail

Voicemail is a feature provided by your cell service so that when you don't answer your phone, the person calling you can leave you a message that you can check later. It's like having an answering machine that you can check from anywhere from your phone.

When you have a voicemail, it will usually show up as a notification at the top of your phone, and you can even drag down from there to listen to your voicemail if you don't want to have to go to your phone app (as shown in figure 11.7).

Figure 11.6

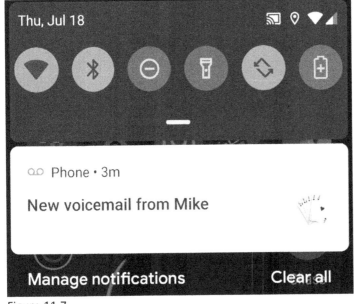

Figure 11.7

In order to use the voicemail feature on your phone, you will need to set it up with a password and a greeting. You can use a personalized greeting or stick with the default robotic sounding one provided by your cell provider. To set up your voicemail, you will need to call into it first and follow the prompts.

There are two ways to check your voicemail from your phone, and if you have the option for what is known as Visual Voicemail (for Verizon, that is) then that is the option you want to go with. Most of the time they have a free version, which is just fine, or a pay-for version which gives you other options you probably don't need.

The Visual Voicemail version shows your email as messages within your phone app while the other method requires you to call into your voicemail, enter a password, and then follow the prompts to listen to your message (and takes much longer to do).

Figure 11.8 shows an example of Visual Voicemail that can be accessed by going to the Voicemail section of your phone app. As you can see, all the voicemails are listed as if they were email messages with their dates and times and all you need to do is tap on the play button to listen to them. Another thing to notice is that one of the messages is bold because it's new and also the Voicemail icon has a notification (number 1) above it.

To play a voicemail simply tap on it, and it will open up and play right on your phone without having to call into your voicemail (figure 11.9). You can also stop, pause, rewind, and fast forward as needed. You can also do things such as send a text message or call the person back right from the voicemail player.

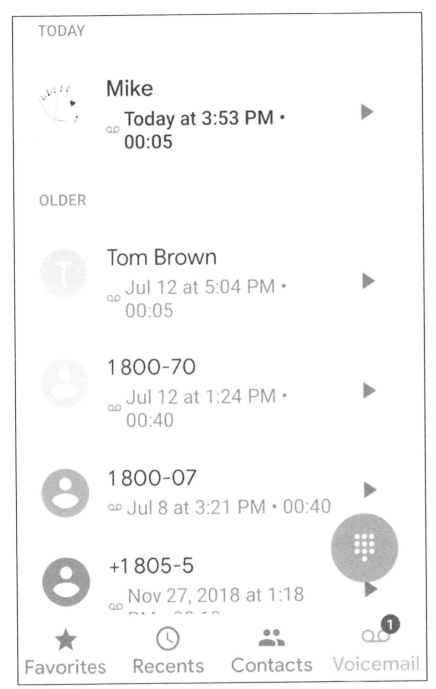

Figure 11.8

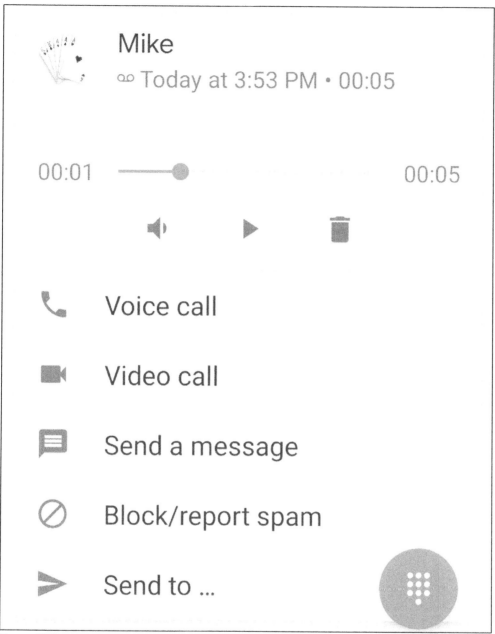

Figure 11.9

If you don't have Visual Voicemail or something similar to it, then you will need to dial into your voicemail, enter your password, and go through all the prompts to listen to each voicemail. This can be time consuming and frustrating if you are in a hurry or get a lot of voicemails. I would check and see if you have this type of option that you can add to your account.

One tip that comes in handy is that when you are receiving a call, you have the option to manually send it right to voicemail so you don't have to listen to it ring a bunch of times before going to voicemail on its own. If you look at figure 11.10, you will see the option to swipe up to answer or swipe down to reject. When you swipe down, it doesn't hang up on the person, but rather connects them right to your voicemail.

So, if your voicemail automatically picks up after five rings and you get a call, you can send it to your voicemail after one or two rings, or however fast you can swipe down to reject the call. Keep in mind that the person on the other end will hear the same number of rings, so if you reject the call after two rings, they might know you are sending them to your voicemail because normally most phones will ring more than twice before going to voicemail.

You can also reject the call and have your phone open your text messaging app so you can send them a text message instead by using the *Reply* message option.

Figure 11.10

Making Video Calls

If you have used a program such as Skype on your computer, then you have some experience making video calls. A video call is when you can talk to someone and see them at the same time and they can see you as well. Apple has been doing this for a long time between iPhone users with their Facetime app. When you go to a contact, you will see the video option, but it may or may not be configured to

181

work, so you may see one of several options for the video call mode such as invite, set up, or video.

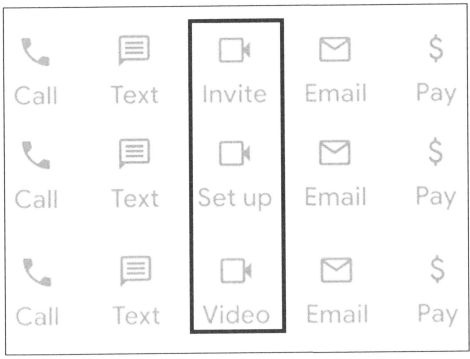

Figure 11.11

If you see *Invite,* then it should send that person an invitation for a video call. If their phone is configured correctly, it should connect, and you should see them on your screen. If it says *Set up,* then you most likely need to do some kind of setup on your end to get things going. (It should walk you through the process.) Finally, if it says *Video,* then things should be ready to go and you can initiate a video call by tapping on the icon.

After the person on the other end accepts the video call, you will see them on your screen, and also see yourself in a smaller box or circle at the bottom of the screen so you can see how they are seeing you on their end (as shown in figure 11.12).

Figure 11.12

Your phone might use the Google Duo app for video calls, or it might use something different such as the manufacturer's built in video calling app. There are also many third party apps such as WhatsApp that you can download and use for video calls.

Airplane Mode

I wanted to quickly mention Airplane Mode because you have most likely heard of it, but may not know what it is or how to use it. When you fly on commercial airlines, they require that you turn off the cellular data connection on your phone for safety reasons. The rumor has always been that your cellphone signal can interfere with the airplane's systems but it's most likely an old wives tale.

When smartphones first came out, airlines made you turn off your phone completely while the flight was in progress, but now that we have airplane mode, we can continue to use our phones for things that don't use cellular connections such as playing games or reading books that we have previously downloaded to our phones.

Airplane mode works by shutting off the cellular connection to your phone while leaving everything else working, including Wi-Fi, so you can't make or receive calls or use your data for web browsing while in airplane mode.

The easiest way to get into airplane mode is to swipe down from your notification area at the top of the phone. From there, you should see an option for airplane mode that you can tap to turn on as well as turn off (figure 11.13). You can also get to it from your Settings app if you go to Network & internet (or maybe just Network) and enable it from there (figure 11.14).

Figure 11.13

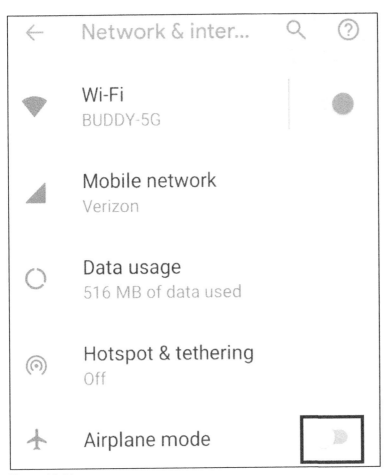

Figure 11.14

Chapter 12 – Texting

Thanks to cellphones, we have one less reason to talk to people face to face, and thanks to smartphones and their ability to send text messages as fast as you can type them, we have yet another reason not to talk to others face to face!

Text messages are just like they sound, messages composed of text that we can send to each other via our smartphones. This way you can send a person or a group of people some information instantly without having to call them. And if they don't read the message right away, it will be there waiting for them to read later.

Sending and receiving text messages is very simple, but there are still a lot of parts to a text message that you should know about. Figure 12.1 shows a text message thread with the important parts labeled. By the way, a *thread* is a continuous group of related back and forth replies between two or more people. You can also have threads in email messages.

Most of the parts of the text message should be pretty self-explanatory, but I will go over them just to make sure you know what they all mean.

- **Text message recipient** – This shows the name of the person you are texting. It's taken from your contacts, so if the person is not in your contacts, it will just show the phone number.

- **Call recipient** – This gives you the option to call the person you are texting right from the text message box.

- **Options** – I will be going over the text messaging app options later in this chapter.

- **Recipient's reply** – What the person on the other end has sent to you.

- **Sender's reply** – What you have sent to the person you are texting.

- **Auto reply choices** – Some phones will give you suggestions for replies that you can tap on to have them sent automatically without you having to type them.

- **Attachments** – You can attach things like pictures to your messages, and I will be going over this process later in this chapter.

- **Compose message box** – This is where you type the text you want to send.

- **Send button** – Once you type your text you tap on the Send button to have it sent to the other person.

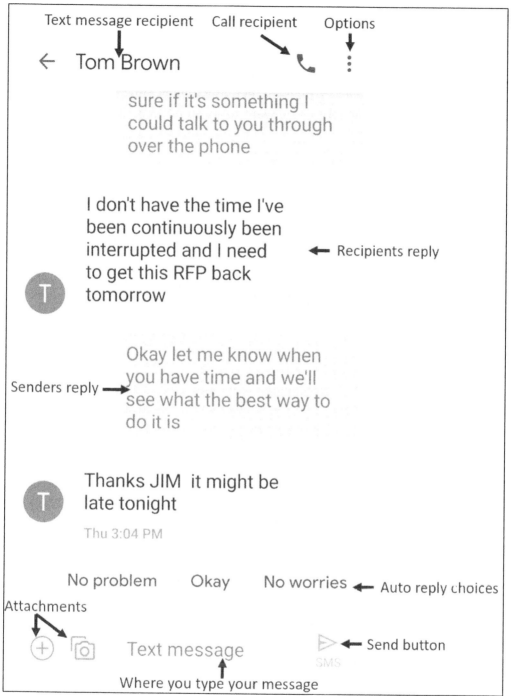

Figure 12.1

Keep in mind that if you are not using the default Android text messaging app or if your phone comes with its own custom app then things might look a little different.

Composing a Text Message

Sending a text message is a very simple process, and you will use some of the same techniques I already went through in other chapters. The first thing you need to do is find your text messaging app, which will look something like the image below. Your phone will come with a built-in text messaging app, but you can also download different texting apps if you want to try something different or that has additional features.

Once you are in your text messaging app, you need to decide who you are going to text. If you have any current text messages that you have sent or received, you will see them there, otherwise it will be blank. You can either text someone from your contacts, or manually enter a phone number.

Figure 12.2 shows a typical Android messaging app with some existing messages. As you can see, there are some individual text messages as well as some group text messages, indicated by two or more pictures to the left of the message. (I will be getting into group texting later in this chapter.) Messages to contacts that you have contact images for will show up with their contact picture next to the message.

To the right of each message you will see the day or time of the **last** message, so don't go thinking it's when the text conversation started. If a text message contains a picture, then it may or may not show up to the right of the message depending on what app you are using.

At the top of the screen, there is a search option that you can use to find text messages with specific words or phrases in them. Then, of course, we have the three vertical dots, which is used to access the options for the messaging app.

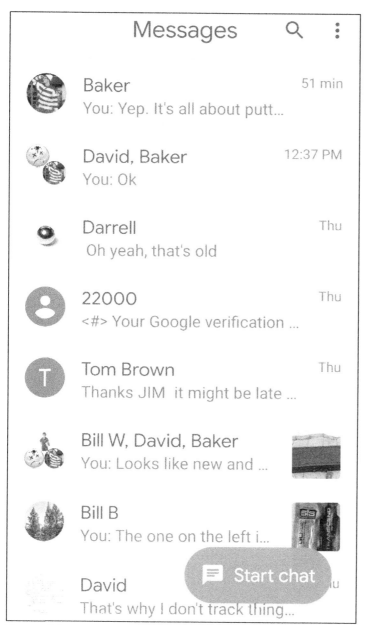

Figure 12.2

To send a text you can simply tap on an existing text message to continue the conversation or tap on *Start chat* to start a new text message. Then you will be brought to your contact list where you can select a contact, or you can choose the number keypad to enter in a phone number manually (figure 12.3). Keep in mind that you will need to enter the area code for any numbers you would have to do that to in order to call.

When you text someone from your contacts, make sure you choose their mobile number and not something like their home or office number if you have that information stored for that contact as well. Otherwise, they won't get the message, and your phone may not know that it's not a mobile number and that it didn't go through.

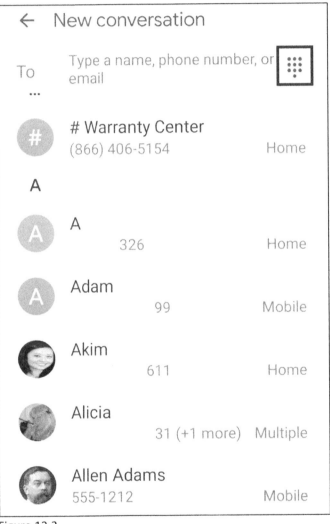

Figure 12.3

Then, once you choose a contact or enter a phone number, all you need to do is type in your message and send it!

Adding Emoji's

You have the option to add what they call emojis to your messages to help portray the emotion you are trying to get across with your message. These can come in the form of smiley faces, sad faces, angry faces, and so on. They can also be things like food and animals if that's what will help get your point across. Figure 12.4 shows just a small sample of the types of emojis you can add to a text message.

Figure 12.4

The way you add emojis will vary from phone to phone. Your phone might have a dedicated emoji button, or you might even have an emoji key on your keyboard or within the box that contains the message itself. Figure 12.5 shows how you might get to your emojis. I first tap on the + button to bring up a list of what type of items I can add to my text message. Then I tap on the emoji option in the first row, which brings me to the available emojis that I have on my phone. As you can see in figure12.6, there are several categories to choose from, so you can have some fun playing around to see what options you have.

Once you find the one you like, simply tap on it to add it to the text message. Keep in mind that it will add the emoji at the point where your cursor is, so make sure it's at the end of the sentence if that's where you plan on adding your emoji. If you change your mind and don't want to add the emoji, then you can remove it just like if you were deleting text.

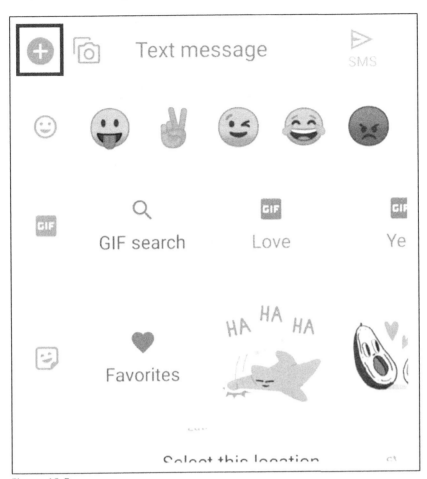

Figure 12.5

Figure 12.6 shows the various emoji categories that I can choose from and also shows how some phones will have the insert emoji option in the same place as where you type your message.

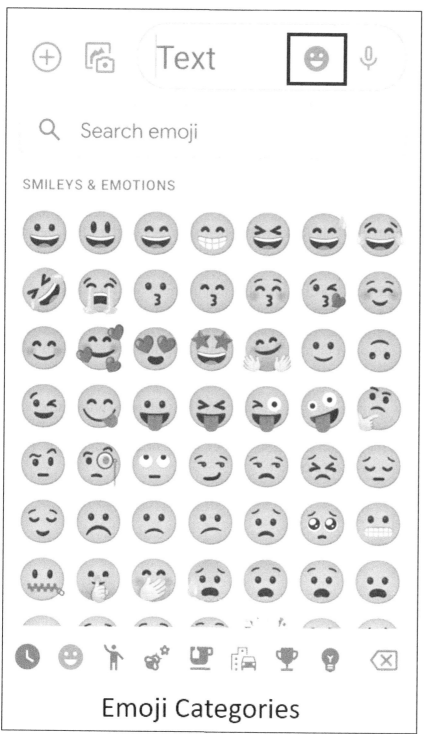

Figure 12.6

Picture and Video Attachments

As you probably know, people like to share the things they are doing with other people whenever possible. If you are a social media user, then you have most likely seen this firsthand! Thanks to smartphones, we can now share our pictures and videos with others by sending them directly to their phones.

Attaching pictures to text messages is a very common practice, and once you figure out how easy it is to do, you will be using this feature all the time. You can do the same with videos, but it's not quite as effective because in order to send these larger types of attachments, they must be compressed to reduce their size and then the quality tends to suffer quite a bit. Pictures will be reduced as well, but for the most part they still look great when they get to the other person's phone.

I will now go over the process of attaching a picture to a text message, and just keep in mind that it's the same for attaching a video, so if you can do one, you should be able to do the other. Once again, my examples may not look exactly the same as what you see on your phone when you go to attach a picture to a text message.

The first thing you want to do is open your text messaging app and select the person you want to send the picture to, just like if you were sending a regular text message. Then look for an icon that looks like a camera or something similar that would indicate you can attach a picture.

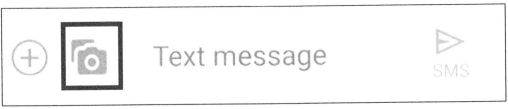

Figure 12.7

Once you choose the insert picture icon, you should have different options for how you want to attach a picture to your text message. You should have the option to browse through the pictures that you have on your phone. Figure 12.8 shows that I have the option to attach one of my latest pictures, or I can tap on Gallery to be shown all of the pictures on my phone and then I can choose one from there.

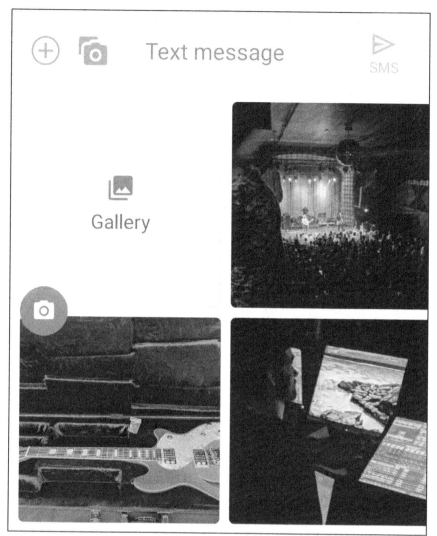

Figure 12.8

Figure 12.9 shows another option you might have on your phone, which is to take a picture on the spot and then have it immediately attached to your text message. This is the option I will use for my example.

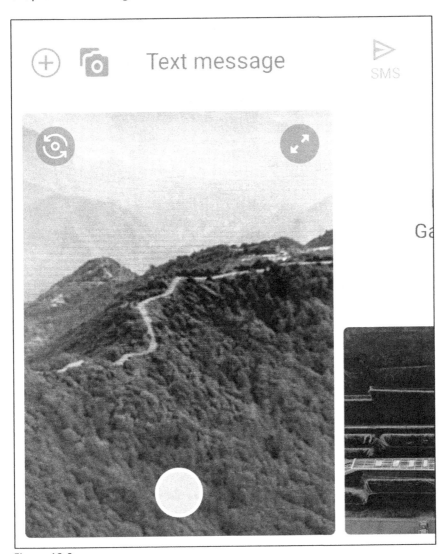

Figure 12.9

After I take the picture, it will be inserted into the text message (as seen in figure 12.10). Then I can add some text to the message to go along with the picture and even add an emoji if I like. Notice the X on the upper right hand corner of the picture. If you tap that, it will remove the picture from the text in case you decided it's not the picture you want to send. Once you click on the send button, then the picture will be sent and it will be too late to remove it. Before you send the picture, you can also add more pictures and send them all at once rather than having to do a bunch of pictures one at a time.

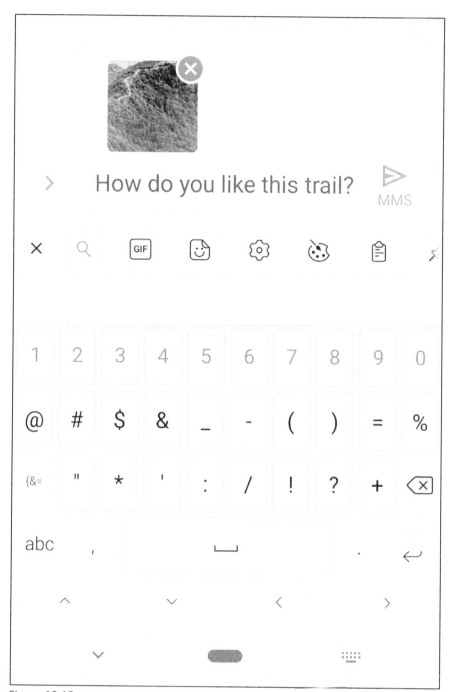

Figure 12.10

Figure 12.11 shows what happens after I tap the send button. It shows a status of sending, and once the picture has been sent, it will change to something like *sent* or just show the time that it went out.

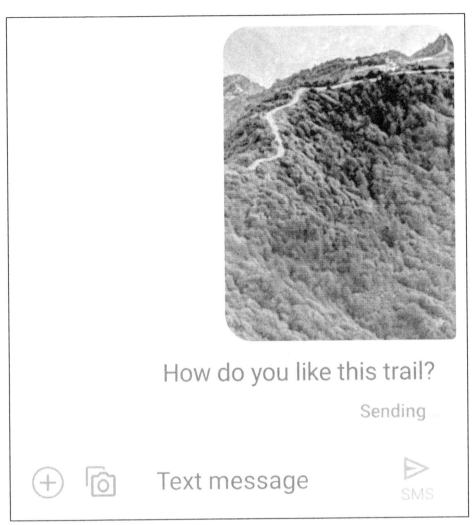

How do you like this trail?

Sending...

(+) [camera icon] Text message ▷ SMS

Figure 12.11

Now that the picture is attached, it will remain in your text message and be part of the message thread (figure 12.12). Tapping on the attached picture will open it up full screen on your phone, and you will also have the option to save the picture so if someone sends you a picture you like, you can save it right to your phone. You can even forward texted pictures to other people if you like. How to do this will vary from phone to phone, but you can try this method: highlight the picture in the message, go to the three vertical dots on the upper right, and choose *Forward*.

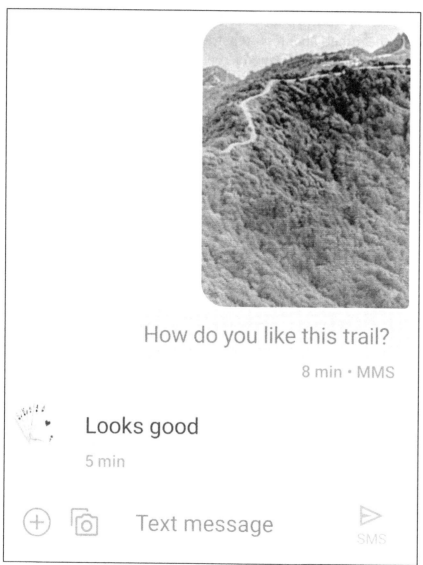

Figure 12.12

Other Types of Attachments

Pictures and videos are not the only type of attachments you can insert into a text message. If you go to your + button (or whatever your add attachment button might be), you will get some additional options for attachments. These options will vary from phone to phone, but you may see options similar to figure 12.13. For example, if you want to share the contact information you have for someone on your phone with another person, you can choose the *Share a contact* option. Or, if you want to attach a document you might have on your phone, you can use the *Attach file* option.

One fun thing you can do is record your voice and then send it as an attachment for the person on the other end to listen to by using the *Touch and hold* microphone option. You might have your microphone option listed after tapping the + button or it might be in the text message box as seen in figure 12.13. (I will be going over text to speech and how it works in Chapter 16.)

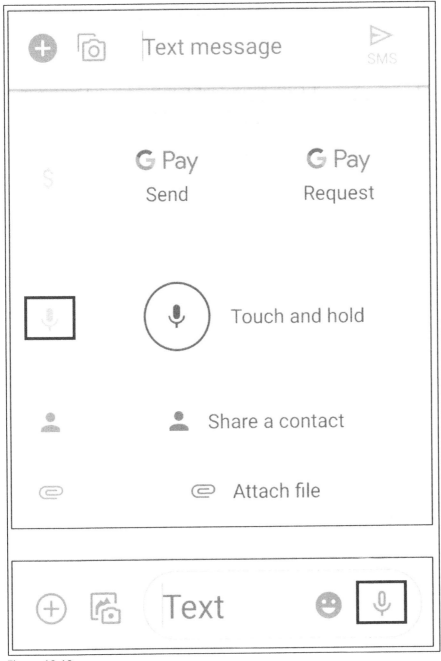

Figure 12.13

Sending Text Messages to Multiple People (groups)

Many times you will have the need to send the same text message to many people. Rather than sending the same message to each person separately, you can send a group message in which everyone will get the same message all in one text. To do so, simply start a new text message like you normally would, but after you add the first person, go back to the To box and tap in there to start adding the next person. Once you start typing, it should start auto populating the names from there. When you have all of the names of the people you want to send the message to, tap on the arrow (or maybe the word next) to get to the next step.

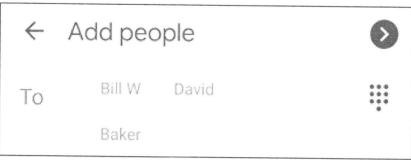

Figure 12.14

You may or may not have the option to give the group of people a name to make it easier for you to keep track of who they are when you see the text in your list of messages. This step is optional, and only you will see the group name.

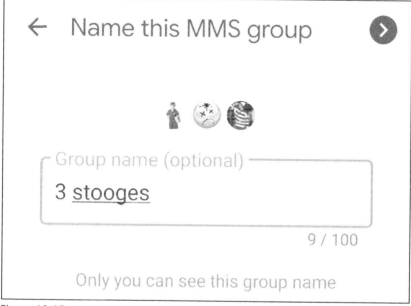

Figure 12.15

Now when you go back to your text message list, you will see that group message with the name you gave it (figure 12.14). If you don't give it a name, then the messaging app will most likely just try and squeeze as many of the recipient names on the list as possible. One important thing to keep in mind when using group messages is that everyone will see everyone else on the list of people you sent it to. If they don't have a certain person's contact information on their phone, then they will just see the phone number and not the person's name.

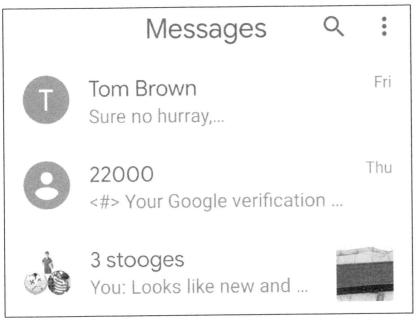

Figure 12.16

Text Messaging App Settings
Your text messaging app will have a variety of settings that you can configure to make the app work the way you like. Not all apps will have the exact same settings, and not all of the settings are anything you would ever change, so therefore I will just go over the ones I feel are important to know about. To get to your settings, you should be on the main screen of the app (not in a specific text message), then tap on the three vertical dots at the upper right and then choose *Settings* from there.

Figure 12.17 shows some of the typical types of settings you might see in your messaging app. Here you can change some of the chat features such as to have your phone use wireless for data over cellular when available, or to have your phone automatically try to resend a message if it fails to be sent the first time. If you want to change the default text messaging app to a different one that you have installed, you can do so here. You don't need to stick with the messaging app that came with your phone if you don't like it, and you can download others from the Play Store.

I went over notifications in Chapter 4, but you can customize them here specifically for the app. You can also do things such as change the resolution for attached photos and enable pinch to zoom for your messages to make them easier to read.

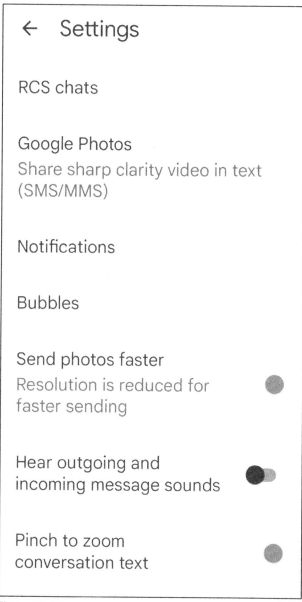

Figure 12.17

You should also have some advanced options if you really want to fine-tune your messaging app. Here you can do things such as enable or disable automatic downloading of your text message (even though I don't know why you *wouldn't* want it to be automatic). You can do the same when roaming as well, but it might be a good idea to leave it disabled to avoid extra charges on your phone bill.

If you want to know when a message was delivered to someone, then you can enable the delivery report option, but I don't know how accurate that really is or if you can trust it. The Wireless emergency alerts setting is used to send automatic alerts to your phone to notify you of things such as fires or floods in your area and even Amber Alerts close to you.

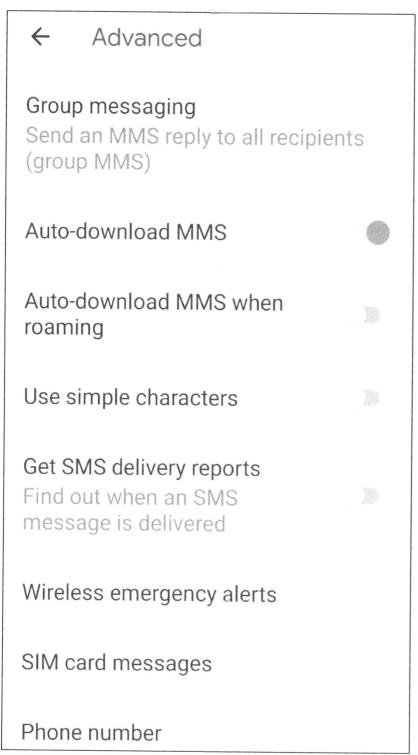

Figure 12.18

Chapter 13 – Configuring Your Phone's Settings

So far you have seen that various apps have settings that you can get into so you can configure how they work to improve your experience with that app. Just like with apps, your phone itself has many options that you can adjust to improve the overall experience you have with your phone.

In this chapter, I will be discussing the more commonly used settings that you might want to change to make your phone work the way you want it to. Not all phones will have the exact same settings, be located in the same place, or be named exactly the same (unfortunately). To get to your settings, look for an app called Settings with a gear icon, which you should have used by now.

Figure 13.1 shows what a typical settings screen might look like. As you can see, it's broken down into several categories that you can then go into to change settings related to that category.

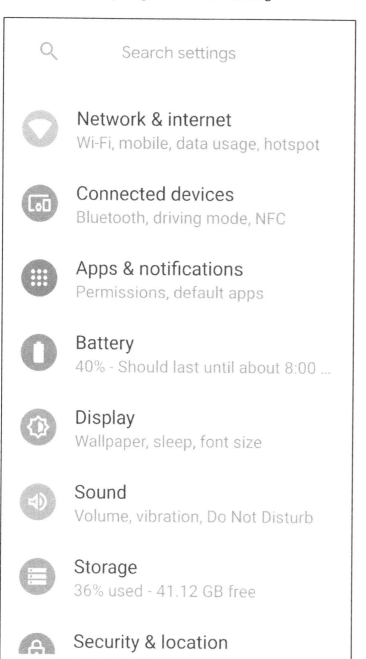

Figure 13.1

Some of these settings I have already gone over in other chapters, so now I will go over some additional settings that I feel you should be aware of.

Connected devices
Your smartphone not only has wireless capabilities but will also allow you to connect to Bluetooth devices such as wireless headsets so you can legally talk on your phone while driving your car.

When you first configure a device to be connected to your phone, you will have to pair that device to your phone so they both know about each other. Most of the time you can simply turn on the device you want to connect to your phone, then go into the Connected devices section, have your phone find the device, and then connect to it. After that, it should remember that device and automatically reconnect to it the next time you have the device powered on and in range of your phone.

Battery Settings
Today's cell phone batteries last quite a long time, but you may find that yours doesn't seem to be lasting as long as it should and might be wondering why. Or you may be simply wondering what apps are using up most of your battery just because you are curious.

If you go to the battery settings for your phone, you can find a wide variety of information related to your battery's performance and what apps are affecting how long it lasts. Feel free to poke around in here to get an idea of how your battery is performing and how long a typical charge will last you based on your app usage.

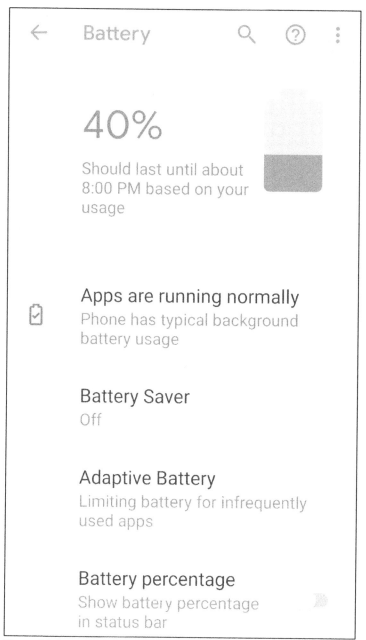

Figure 13.2

Display Settings

I went over some of the display settings when I discussed how to change your background wallpaper image back in Chapter 3, but there is much more you can do from the display setting than just changing your wallpaper.

One of the more commonly changed settings that is done from here is to change the brightness level of your screen. Many phones will have it set to automatic, but if it's not bright enough for you or too bright for you, then you can adjust the setting here.

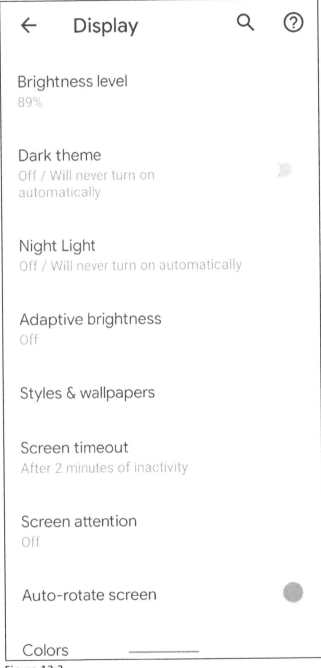

Figure 13.3

Another common setting to change is the sleep timer for your screen. This is the amount of time before your screen will turn itself off if you don't touch your phone. Many times it's set to 30 seconds, which many people think is too quick, so I like to change mine to two minutes. This means my screen will stay on for two minutes if I don't touch it. Of course, if you are constantly using your phone, your screen will not turn off at all (unless your battery dies!).

I'm sure you have noticed that when you turn your phone from vertical to horizontal (or vice versa), the screen orientation changes with it. Most people prefer this to happen, but if you are the type that doesn't, then you can disable that feature here. Many times I have experienced a situation where this gets disabled on its own and you need to go into the display settings to re-enable it.

Sound Settings

Your phone will have various sounds that it makes for various functions such as phone calls and the alarm clock, etc. From the sound options you can change setting such as the actual sounds themselves, the volume of these sounds, ringtones, and more. Figure 13.4 shows how you can adjust the volume of your media (music, games, and videos), call volume, ringtone, and alarm clock, as well as turn on the vibration feature of your phone so not only will it make a sound, but also vibrate so you can feel it in case you can't hear it.

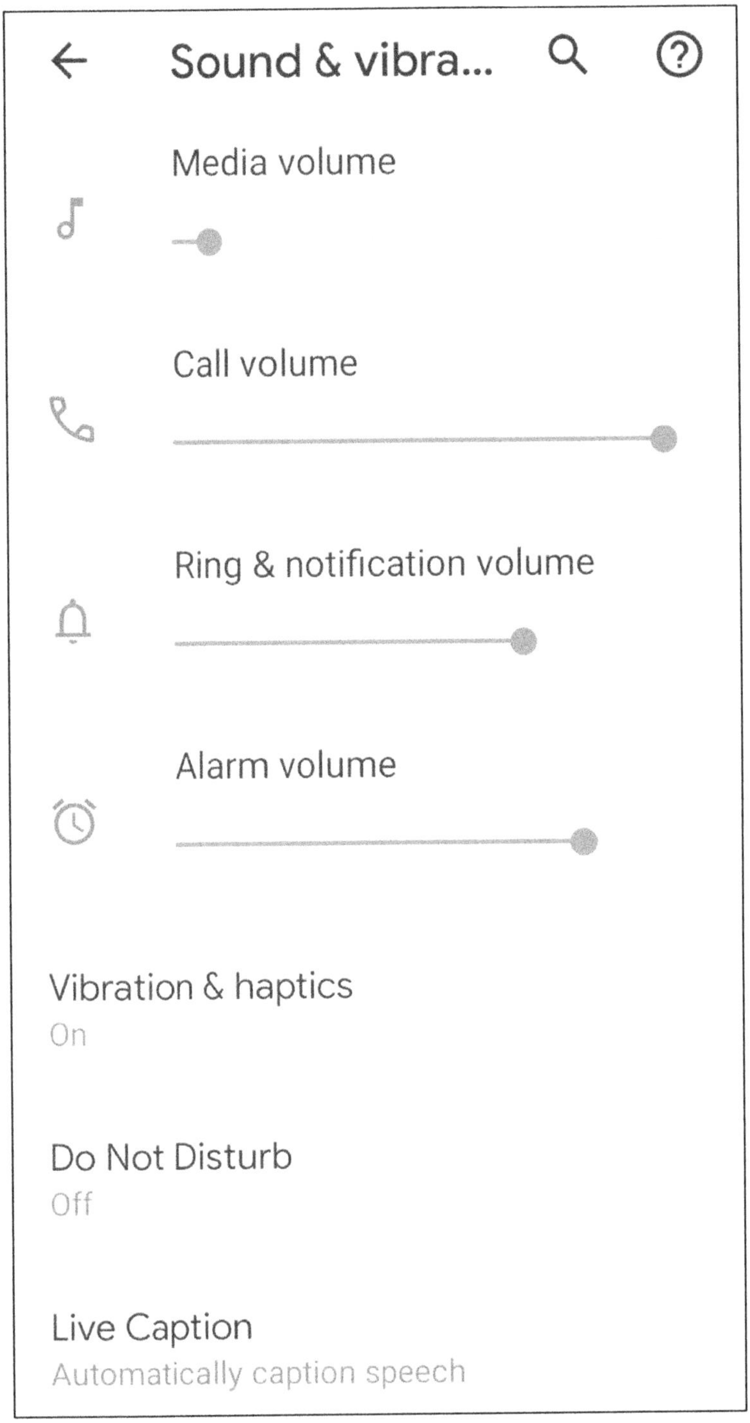

Figure 13.4

Figure 13.5 shows some more advanced options such as how you can change your ringtone, notification, and alarm clock sounds from the defaults. Most phones will have a variety of built-in sounds to choose from. To change your ringtone, simply go to this section, sample some ringtones, and choose the one you want—that's all there is to it!

If you are the type who doesn't like to hear the beep while dialing numbers on your phone or have it vibrate each time you press a key, then you can adjust those settings here as well. I prefer to keep my phone as quiet as possible as to not disturb others and myself!

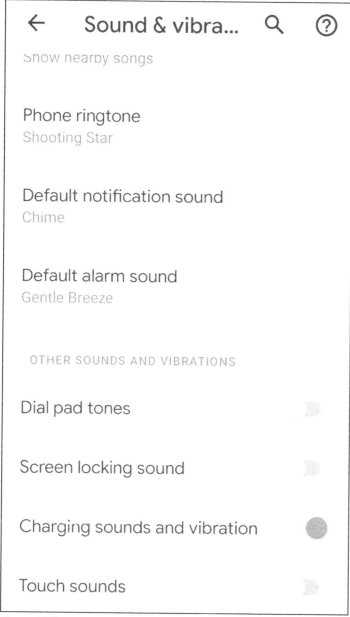

Figure 13.5

Accessibility Settings
If you have certain limitations that make it difficult for you to perform certain tasks on your smartphone, then you might want to take a look at the accessibility settings on your device to see if there are some options there that can make things easier for you.

Here you will find various options that you can configure to make things easier on you to make your smartphone experience more enjoyable. For example, you can have your phone read text to you just by tapping on it, or have your phone communicate with you verbally so you don't have to look at the screen as often.

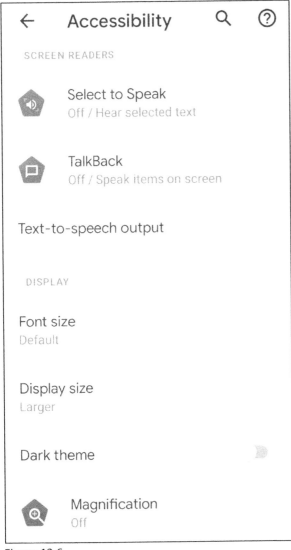

Figure 13.6

If your eyes aren't what they used to be, then you can do things like increase the font and display size, or enable the magnifying glass option, which will let you zoom in on parts of your screen to help you better see that section. There are other options to change colors and contrast as well, making things easier to see and read.

If this is something you think you might be interested in using, then take some time to see what options you have available to you on your phone and play around with them. You can always change things back if you end up turning on a feature and not liking it.

Finding Information About Your Phone

At some point during your smartphone experience, there will come a time when you need to find out some information about your phone. Most likely it will be because you are on the phone with support and they need to either verify your phone model or build model, or want to check its IMEI or MEID number, which is used for activation purposes.

Within the various settings you should have one that is labeled *About phone* or something similar. Here you can find the information I have mentioned above, as well as other things such as your phone number (in case you forgot!), networking settings, your Gmail address, Android operating system version, and so on.

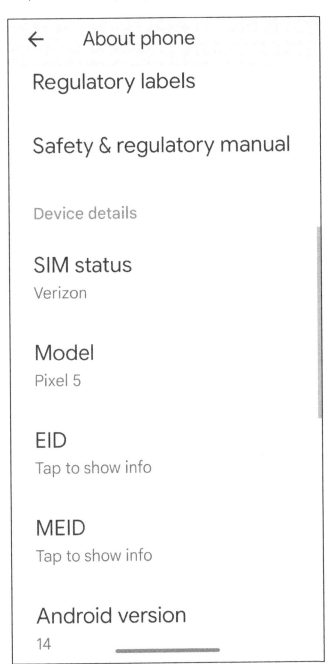

←   About phone

Regulatory labels

Safety & regulatory manual

Device details

SIM status
Verizon

Model
Pixel 5

EID
Tap to show info

MEID
Tap to show info

Android version
14

Figure 13.7

One helpful setting you might want to change from here is to enter your emergency contact information. That way, if you are in trouble or in an accident, the people trying to help you can get this information from your phone even if it's locked.

Chapter 14 – Navigation

In my opinion, one of the best things about smartphones is the navigation feature, which will help get you to almost anywhere you need to go. This eliminates the need to have a GPS in your car (even though many cars still come equipped with them). The navigation feature can be used while you are walking, and also can be used for public bus schedules in case that is how you are traveling.

Google Maps

There are many ways you can use your phone to navigate, but the most popular for Android users and even some iPhone users is the Google Maps app. This will come installed on your phone, so there is no need to go out and download it first in order to use it.

The Google Maps interface is pretty easy to use. All you need to do is type in the address of where you want to go, and it will find the directions for you. It does this by using your phone's built in GPS (global positioning satellite) to find your current location and then find the best route to your destination.

For example, if I type in Boundary Bay Brewery in Bellingham WA into the search box, I will get the results shown in figure 14.1. Notice how it brings up suggestions related to what I'm searching for in case I need help trying to find the right place. If the right choice is shown in the suggestion area, then I can simply tap one of those. Otherwise, I can just choose from the search results after I type it in.

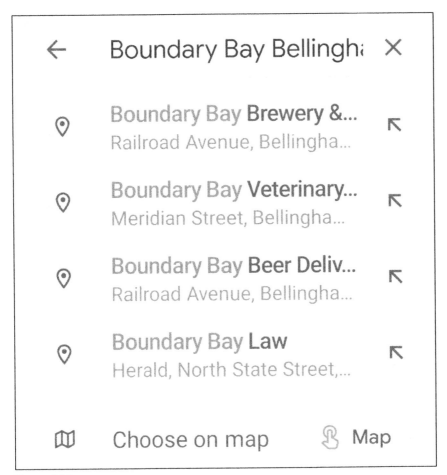

Figure 14.1

Figure 14.2 shows me the results of my choice. As you can see, there is all sorts of information provided to you from the app such as its location on the map, nearby places, photos, reviews, an option to call the location right from the app, and, of course, the button to tap on to get directions.

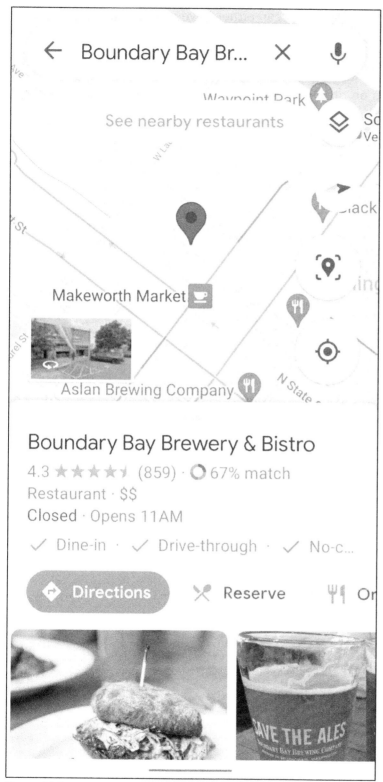

Figure 14.2

When I tap on the *Directions* button I can either have it find directions from the location I happen to be at the moment or type in a starting point in case I want to find directions from a specific place. Also notice that you have options to choose how you are getting there such as driving, taking the bus, walking, riding your bike, and so on. There is also an option to check traffic in your area, which is always helpful!

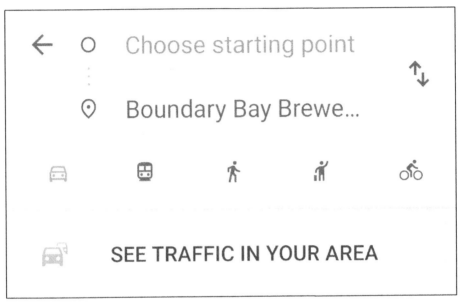

Figure 14.3

Figure 14.4 shows the route Google Maps found to get me to my destination as well as the distance and how long it will take using various modes of transportation. It will also show traffic in the surrounding areas with green meaning no traffic, yellow and orange meaning moderate traffic, and red meaning heavy traffic.

Figure 14.4

If I tap on the *Steps & more* button, it will show me step by step or turn by turn directions that you can use if you are more comfortable navigating that way. If there is more than one route available, then it will also show you other suggested routes with their distance and time information.

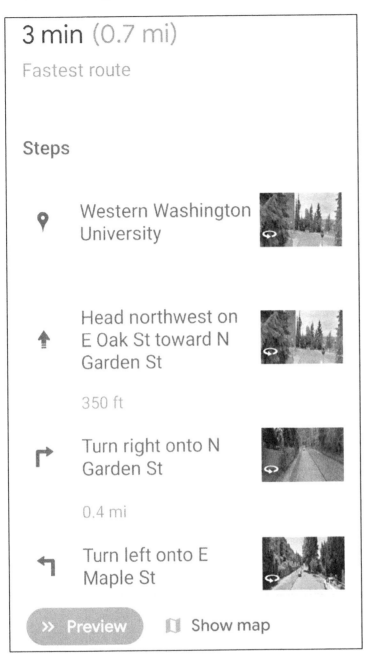

Figure 14.5

One nice feature of Google Maps is the ability to change the map type. If you don't like the default street map, then you can change it to satellite mode by tapping the layer button at the top right corner of the map (as seen in figure 14.6). Then you can choose the satellite layer (14.7), and the results are shown in figure 14.8.

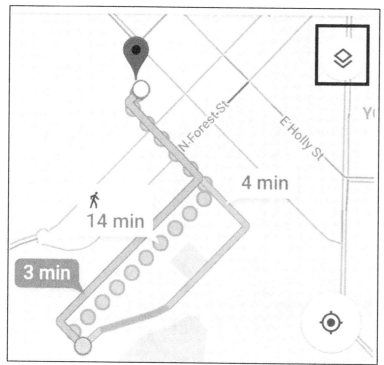

Figure 14.6

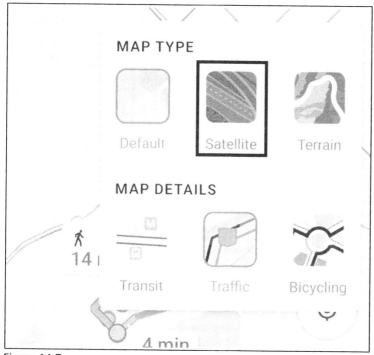

Figure 14.7

Figure 14.8

When you are searching for directions from a starting point that is not where you are physically located, you can only get the directions and can't actually navigate there since you are not there to begin with. When you use "your location" as the starting point, then you will have the option to begin the navigation process and will have a button labeled *Start* at the bottom of the screen.

Figure 14.9

Once you start your navigation, the map will move on the screen, showing where you are at all times, and you will hear a voice telling you what your next turn will be and how far away it is.

I also wanted to mention some of the other cool things you can do from Google apps such as look at restaurant menus, share the place with others, look at photos, order food, get business hours, and so on. This information will vary depending on what type of place it is. For example, you won't get the menu option if you are looking for a shoe store.

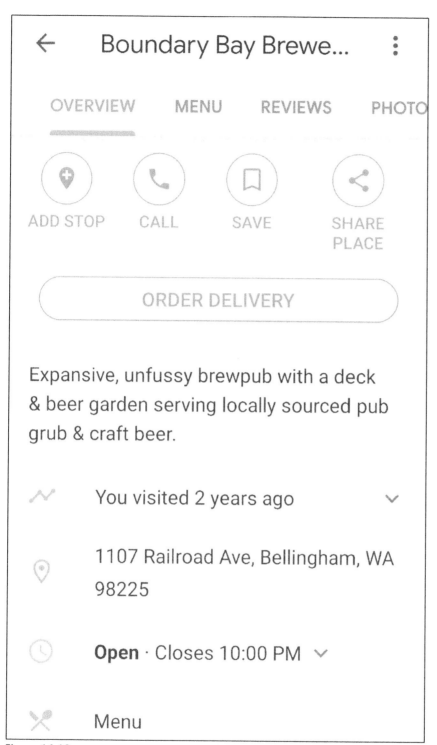

Figure 14.10

Searching for Services

Not only can you use your navigation app to find directions, you can also use it to find goods and services such as restaurants, hotels, gas stations, and anything else you can think of.

This can be done from your current location, or you can search for a location and then find services around there. For example, I can type in "shoe stores" right in the search box and it will start to auto fill with results.

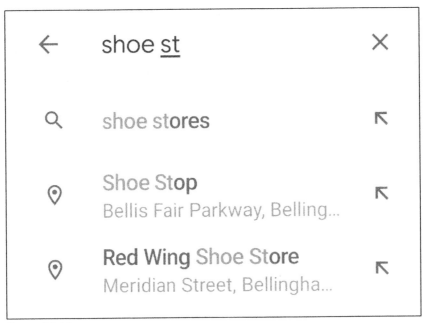

Figure 14.11

Once the search is complete, I will be shown the results on the map that you can tap on to expand, as well as a listing of shoe stores close to that area that I can tap on to get more information about (as you can see in figure 14.12). Notice how I get additional options to filter my results such as stores that are open now or have the best reviews.

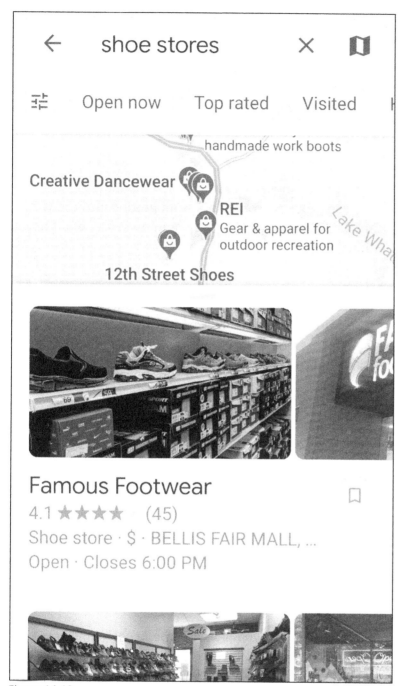

Figure 14.12

Overall, you can use Google Maps or the navigation map of your choice to pretty much find anything you need and find out how to get there.

Chapter 15 – Security

Just like with computers, smartphones are becoming a valuable target for criminals looking to steal from you. Whether it be physically or virtually, you need to make sure that your phone is as secure as it can be to keep yourself and your information protected.

In this chapter, I will be discussing various things you can (or should) do to increase your safety level and to prevent your personal data from getting into the wrong hands. The most important way to stay safe is to use common sense, but it never hurts to have some additional help!

Locking Your Phone
When it comes to security, there are two types you should be concerned with, and they are physical security and virtual security. Physical security is preventing your phone from being physically accessed while virtual security is preventing your information from being compromised from the phone itself.

One way to physically protect your phone is to use a lock screen that prevents others from getting on your phone in case they get their hands on it. Normally when you turn your phone on or even turn on the screen after it goes off, all you need to do is swipe and you have full access to your phone. Of course, this is the fastest way to get back on your phone after being off it, but it's definitely not the most secure.

What you should do is set up either a passcode (PIN), password, unlock pattern, fingerprint access, or even facial recognition access if your phone supports it. Your phone may have additional options, but these are the most common. These options can be found under *Security & privacy* within the Android settings.

A *passcode* is simply a numeric code that you have to type in each time you start up your phone or turn the screen on. Think of it as being the same as how you access the ATM with your PIN number. A *password* is just like it sounds: a word that you have to type in, just like you would on your home or work computer.

An *unlock pattern* is just like it sounds: you draw a pattern on your phone to unlock it. If you look at figure 15.1, you will see an example of a lock pattern that I set up on my phone. If I were to save this pattern, then each time I went to unlock my phone, I would have to draw this pattern (connect the dots) in order to get into my phone.

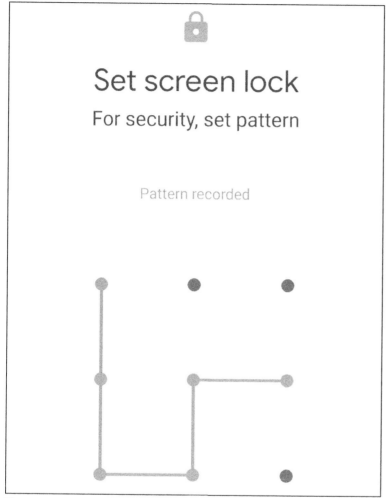

Figure 15.1

Most higher end smartphones have *fingerprint* readers that can be used for a variety of tasks, but one thing you can use it for is to unlock your phone since you are the only one with your fingerprints (hopefully!). Once you configure your phone to recognize your fingerprint, then you can use it to quickly unlock your phone.

Facial recognition is becoming more popular with smartphones and works the same way as fingerprint scanning. You simply position your phone so your camera is facing your face, and once it recognizes you, it will unlock your phone. You can get to these options from the *Security* section in your Settings.

Fingerprint Scanners

I just mentioned how scanning your fingerprint is a quick way to give you secure access to your phone since your fingerprint is unique to you. But unlocking your phone is not the only thing you can do with your fingerprint.

Many apps have security features built in that utilize your phone's fingerprint scanner to help keep them secure. For example, a photo app might have a feature that requires you to swipe your finger before being able to see your photos, or you might also find that your banking app has an additional security feature you can set up so that you have to swipe your finger to log in along with your password.

Some phones offer additional functionality when you use the fingerprint scanner. For example, the Google Pixel line of phones allow you to scroll down the notification panel by swiping your finger down on the fingerprint sensor. Other phones will let you use it as the shutter button when taking pictures, which makes taking selfies that much easier.

If you decide to use your fingerprint scanner for security, just keep in mind that it's not 100% hacker proof, so if you can use it as an additional security layer, then that is the way to go.

Finding Your Lost Phone

Losing your phone can be a gut wrenching experience, almost as bad as losing your wallet (maybe worse for some people), but if you have the right type of phone with the right type of features enabled, then there may be hope for getting it back.

Smartphones have a location feature that is used to keep track of where you are at all times (scary) and is used for things such as your navigation app to get you from your current location to where you want to be. One good thing about the location feature is that it can be used to help you find your phone if you lose it.

 If you are not into being tracked 24x7, then it's possible to disable the location feature on your phone. This can usually be found in your main phone settings. Keep in mind that this will prevent any apps that use your location from working properly.

Many phones have their own built-in apps that will help you find your phone if you lose it, and Google has their own called **Google Find My Device** that you can install from the Play Store. Once you download the app, all you need to do is sign in with your Google account and you are ready to go. Then, if you lose your phone, or even misplace it, you can go to the Google Find My Device website from any computer or other device and log in with your Google account. You will then be shown on a map where your phone is.
https://www.google.com/android/find?u=0

From there you can have the site ring your device so if it's close by (like under the couch cushion) you can hear it. If you don't have a security method set up on your phone, you can have it sign you out of your account and then lock your phone. If you think your phone is gone for good or stolen, then you can have it remotely erased.

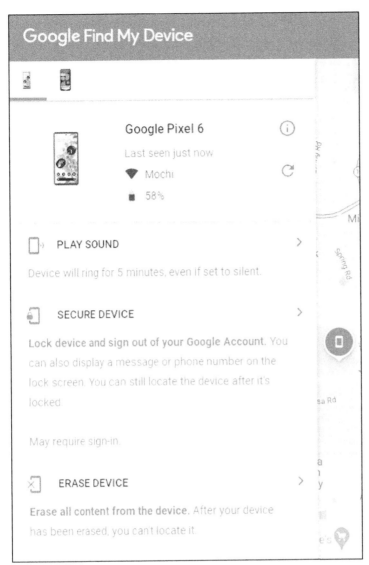

Figure 15.2

If you do use this app, then it's a good idea to bookmark the Google Find My Phone website on a different device (like your home computer) so you will be able to find your phone as fast as possible and not need to worry about what website you need to go to in order to start the process.

Usernames and Passwords

I mentioned passwords earlier in this chapter when talking about locking your phone for security reasons, but now I want to talk about passwords that you use for things like apps and websites. As you probably know, many websites require you to log in with a username and password, and you often have the ability to have your computer or phone store those passwords so you don't need to remember them or type them in each time. Many apps have this option as well.

Figure 15.3 shows an example of a banking app that lets you save your username so you don't need to enter it each time. Notice how it also has a fingerprint sign in option?

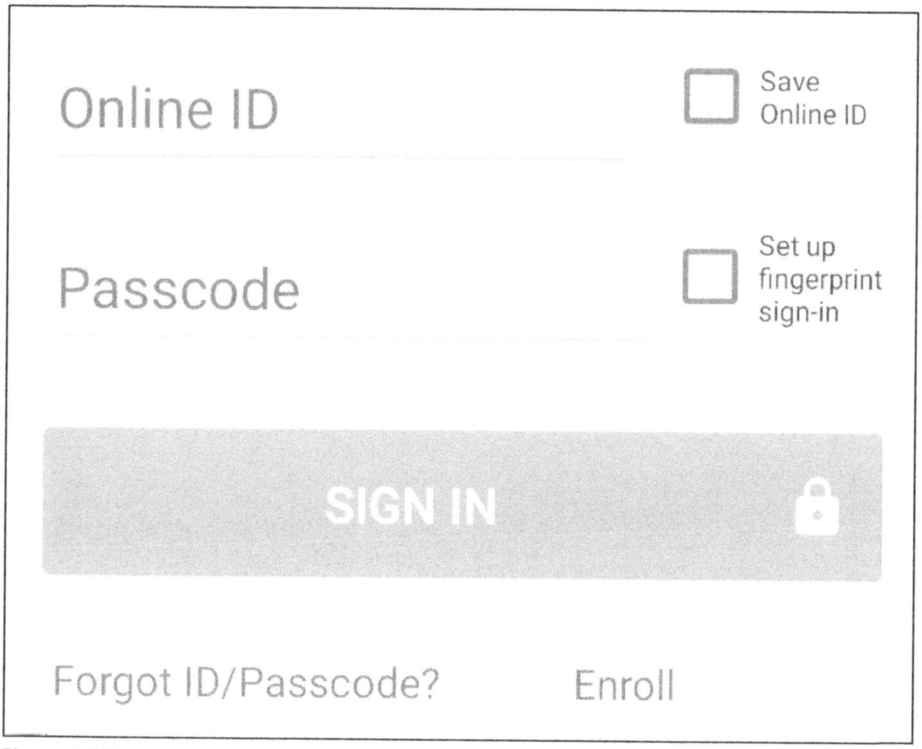

Figure 15.3

Sure, it's tempting to have it save your username, but if you do that and lose your phone, that gives the person who finds it half of your login information to your banking app. Then all they need to do is figure out the password. I would never save any login information for banking apps or sites because of the security risk. You are better off just typing in your username and password each time. (And make sure your password is complex and not something easy to guess.)

Using Third Party Apps

One of the main reasons for having a smartphone is to install apps that can be used to improve the functionality of your phone and justify that high monthly cost. Back in Chapter 5, I went over how to use the Play Store to install apps and what to look for when it came to reviews and the number of downloads.

The Play Store is not the only place you can get your apps from, and if you look online, you can find third party apps available to download that are not officially supported by Google. In fact, you can even find modified versions of the Android operating system that you can use on your phone.

Even though some of these apps look tempting, I do not recommend installing any apps from any source other than the Play Store on your phone itself. The reason for this is that they haven't been checked for things like viruses or bugs that might affect your phone, or (worst case scenario) make your phone stop working altogether.

There is one exception to the rule, and that is downloading apps from the online version of Google Play itself. If you are signed into your Google account on your computer and go to https://play.google.com, you can see the apps installed on your phone and browse for new ones.

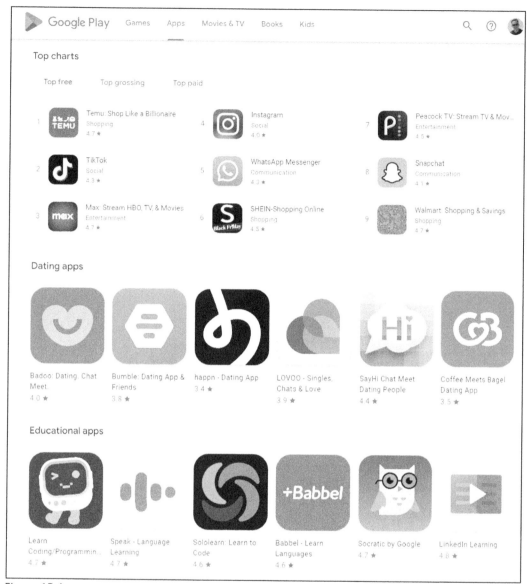

Figure 15.4

If you find an app you like, you can click on Install, and it will actually install the app on your phone right from the website. So, if you are at work and your phone is at home, you can install new apps without even touching your phone!

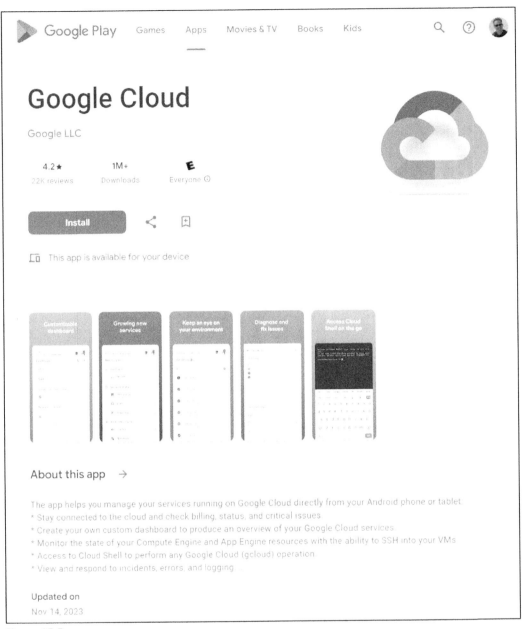

Figure 15.5

Chapter 16 – Tips and Tricks

So, you've finally made it to the last chapter in this book. Hopefully you're a much more proficient Android smartphone user, and actually had some fun learning how to use your phone and discovering what it can do for you!

For this final chapter, I wanted to go over some tips and tricks that I think you might find useful and can use to make your Android experience even that much better. The content of this chapter is not stuff you necessarily *have* to know but definitely won't hurt to know.

Rebooting and Shutting Down Your Phone
Just like with your computer, you can reboot and shut down your phone as needed. You might want to shut down your phone when flying on an airline, or if you don't plan on using it for a while like when on vacation (if you can actually pull that off!).

The main reason you will want to reboot your phone is if it's acting funny or glitchy to see if a restart will clear things up. This is common practice with computers, as you most likely have experienced. When your phone is on for an extended period of time and you are opening and closing apps etc., your phone will tend to slow down and maybe start having issues with apps freezing up on you. Simply rebooting your phone usually clears things up, and I recommend you reboot once a week or so regardless just to keep things optimized.

The process for rebooting your phone is simple. All you need to do is hold down on the button that turns the screen on and off until you get the power menu (like shown in figure 16.1). Then you can tap on Power off (shutdown) or Restart (reboot) and your phone will take care of the rest.

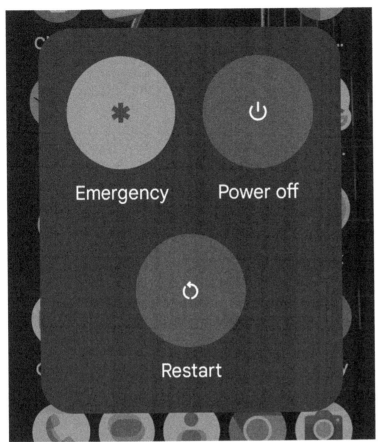

Figure 16.1

Android Updates

If you use Microsoft Windows on your computer at home, then you have most likely had to deal with Windows Updates and how it bugs you to restart your computer so Windows can get things up to date.

The Android Operating system will have updates as well, and you will see notifications on your phone telling you there is an update that needs to be applied. When you see these notification updates, you will need to reboot your phone to have them applied, and sometimes they can take a bit of time to complete. You can expect a typical update to take about ten minutes.

You don't have to do these updates, but it's a good idea to get them done because they usually contain security enhancement and bug fixes that make your phone more secure and run better. Many times they will include new features as well. Just like with any software update, there will be a chance of something going wrong, but it will be rare.

Text to Speech

One of my favorite things about smartphones is the text-to-speech feature. This is where you can talk into your phone and it will translate it to text in the app you are using. So, if you are composing an email or text message, you can simply speak into your phone and it will add the text for you.

Keep in mind that the translation might not be perfect each time, so it's important to speak slowly and clearly while using this feature. You should also proofread the text that it translated from your voice before sending out any text or email messages in case any mistakes were made. Also, try to avoid having noise in the background like the TV on, because you might end up recording what they are saying on the TV, too!

To use the voice to text feature, look for the microphone icon on your keyboard, or within the app itself.

When you tap on the microphone icon, you should hear a beep, and then all you need to do is start talking and it will translate as you speak. When you are finished, simply tap on the microphone icon to have it stop (figure 16.2). If you want to add additional text, then tap on the microphone icon again to start the translation process.

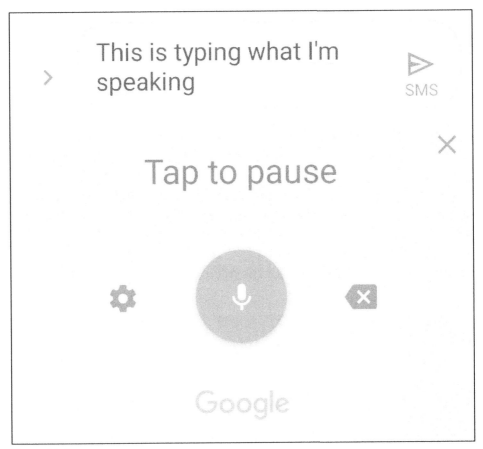

Figure 16.2

To add punctuation to your text, you will need to speak what you actually want it to say. So, if you want to have it type the phrase, "I got a new phone. It's the best thing ever!" you will have to say it like: "I got a new phone period It's the best thing ever exclamation point."

Keep in mind that this feature doesn't always know what you are saying, so when you say the word **too,** it might type it out as *two, to,* or *2,* so be sure to do your proofreading!

Not all apps support the voice-to-text feature, so if you don't see the microphone icon, then that probably means it won't work with that app.

The voice to text feature needs a data\internet connection to do the translation, so you may find that when you are in an area with a weak signal that the feature might not work as well, or even at all. If this happens, don't think it's your phone when it's really just the connection.

Checking your Data Usage

When you do things on your phone like check email, surf the Internet, play certain games, watch YouTube videos, etc., it uses what is called *data* to do so. Data is what your phone sends back and forth between itself and other places such as websites or email servers. Certain activities such as streaming movies or music will use much more data than other activities such as checking your email.

The amount of data that you are allowed to use each month will depend on how your smartphone plan is set up. Some people will have plans with unlimited data, but those plans can be on the expensive side.

Data is usually measured in gigabytes (or GB for short), and a typical plan might allow you to use anywhere from 2 to 8 GB of data per month. If you go over this amount, you are normally charged for the extra GB used. Many plans will have a rollover feature so if you don't use all of your data one month, then you can rollover the leftover data into the next month and have some extra cushion.

I suggest you start with a mid-level plan and see how much data you actually use each month, then adjust your plan up or down as needed. Why pay for 8 GB of data when you only use 4? The amount of data you use will most likely vary each month, so you might want to give it a few months to really see how much you use. You should be able to look at your phone bill to see your totals.

There is a way to check how much data you have been using so far as well as finding out what apps are using how much of your data. Most phones will have this information in the Android Settings under the *Network & internet* section.

Figure 16.3 shows how much data I have used in my billing cycle so far. You might be wondering why it says 90 MB (megabytes) and not GB (gigabytes). That is because I haven't used a GB yet, and a MB is smaller than a GB. In fact, there are 1024 MB in 1 GB. This means I have used about a tenth of a gigabyte.

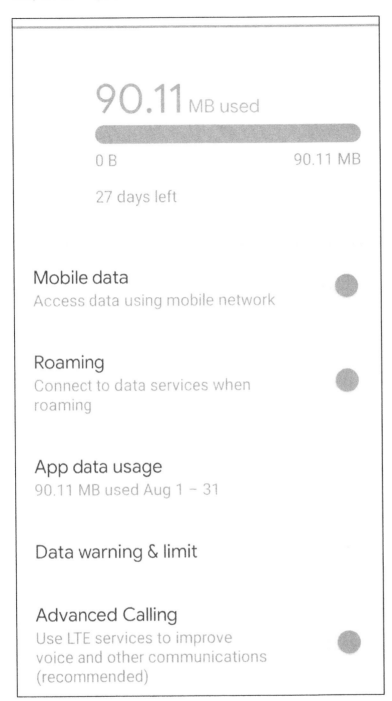

90.11 MB used

0 B   90.11 MB

27 days left

Mobile data
Access data using mobile network

Roaming
Connect to data services when
roaming

App data usage
90.11 MB used Aug 1 – 31

Data warning & limit

Advanced Calling
Use LTE services to improve
voice and other communications
(recommended)

Figure 16.3

To see what apps are using what data, I can tap on *App data usage* and get a listing like the one seen in figure 16.4. Once again, the numbers are in megabytes because if you are using gigabytes of data in your apps, then you are doing something wrong! Also notice how you can change the date range on top to fine-tune the results.

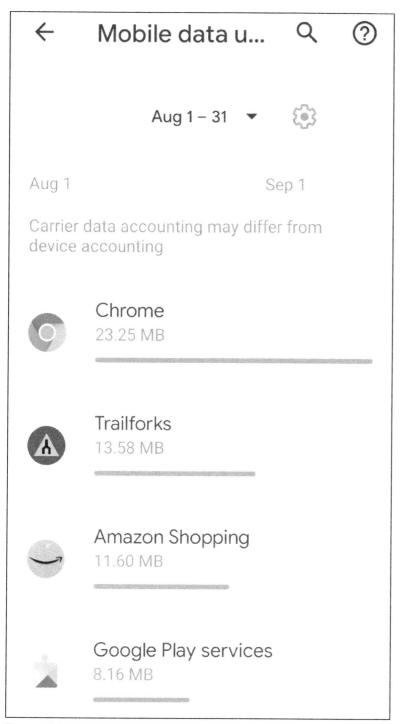

Figure 16.4

If you are doing just the basics with your phone, then you probably won't really need to worry about checking data usage, but if all of a sudden your bill increases, you might want to look into the data settings and see if you can figure out what is causing the increase.

Setting Custom Ring Tones for People
By default, your phone will use one ringtone for all calls, which may be fine for some people, but if you would like to be able to tell who is calling you just by the sound your phone makes, then it's possible to set up different ringtones for different people in your contact list.

To do this, you will want to go to your contacts and pick a person that you want to add a custom ringtone for. From there, you will either tap on the three vertical menu dots and choose *Set ringtone,* or you might have to use the *Edit contact* option.

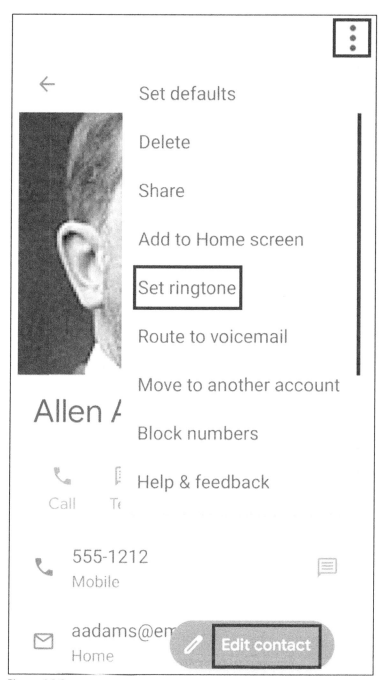

Set defaults

Delete

Share

Add to Home screen

Set ringtone

Route to voicemail

Move to another account

Allen /

Block numbers

Call   Te   Help & feedback

555-1212
Mobile

aadams@em   Edit contact
Home

Figure 16.5

Then you will be taken to the available ringtones that you have on your phone. From there you can sample each one and choose the one you want to use by tapping on Save.

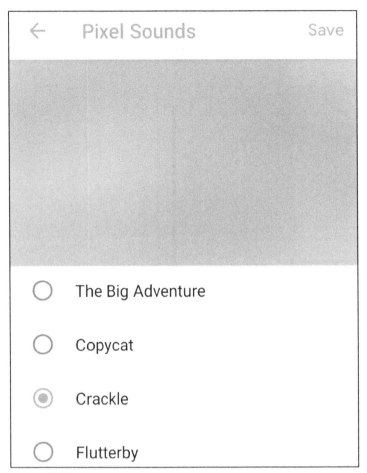

Figure 16.6

If you don't like any of your choices, you can go to the Play Store and download some additional ringtones. Just be careful, because many of these free ringtones will come with ads or other nonsense you don't want on your phone.

Using the Flashlight
Now that you know that smartphones are much more than just phones, it only makes sense that you can use them as flashlights as well... right? Believe it or not, you actually can use your phone as a flashlight.

The way your phone becomes a flashlight is by using the camera flash to provide the actual light. Rather than just flashing it quickly like it would when used to take a picture, it will keep in on instead, allowing you to point the light at whatever you are trying to see.

When smartphones first came out, you needed to download an app in order to use your phone as a flashlight, but now most phones come with a built-in flashlight feature that you can usually access by pulling down from the notification area at the top of your phone (as shown in figure 16.7).

Figure 16.7

If for some reason your phone doesn't come with a built-in flashlight, you can download one for free from the Play Store. I recommend the one called Flashlight

Widget by David Medenjak because it will allow you to put a simple button on your home screen that you just press to turn the flashlight on and off, so you don't have to open an actual app.

Figure 16.8

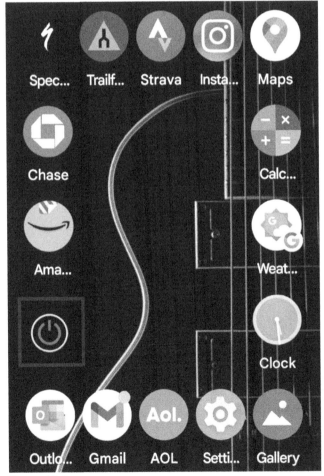

Figure 16.9

Custom Keyboards

One of the most widely used components of a smartphone is the keyboard, so it only makes sense that your keyboard is configured the way you like it so it makes typing on it easy and efficient.

The Google keyboard works just fine for most people, but if you want to mix things up, then you can download and install a different keyboard from the Play Store. Many of these keyboard replacements are very customizable and you can do things like add or remove keys and change the keyboard color or theme.

Figure 16.10 shows an example of the default Google keyboard on my phone when used with the text messaging app.

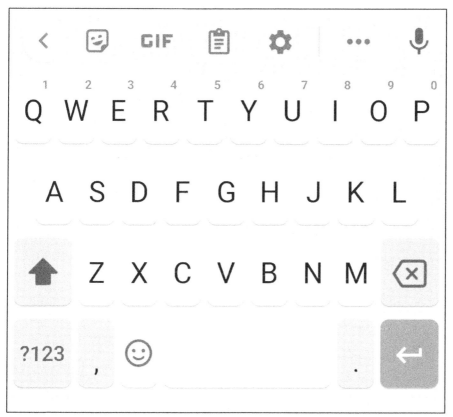

Figure 16.10

Figure 16.11 shows the Swiftkey keyboard that I have installed on my phone in use with the text messaging app. Notice how it has things like dedicated number keys and up, down, left, and right arrow keys. This keyboard will also allow me to change the color theme if I want to mix things up a little.

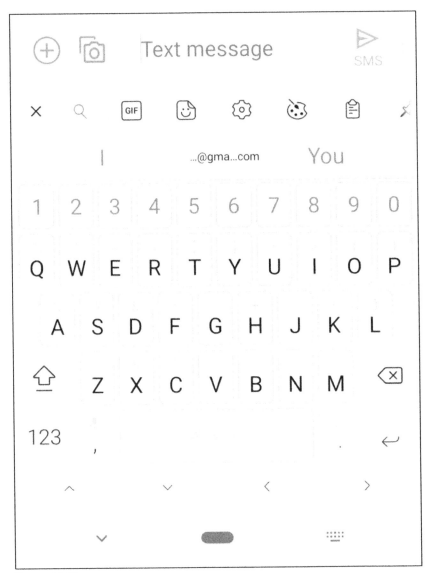

Figure 16.11

Swipe Feature

Speaking of keyboards, Google has a unique feature for typing that is a little on the odd side, but once you get used to it, can come in quite handy. Instead of having to type letters one at a time, you can swipe the word you want to type by starting on the first letter and sliding your finger to the next letter and so on, kind of like if you were connecting the dots. After you touch the last letter of the word you simply release your finger and hope that your phone knew what you meant to "type."

Figure 16.12 shows a sample of how I started to swipe the word "swipe," and my phone knew what I meant and finished it for me so I didn't have to swipe all the way to the last e of the word "swipe."

Figure 16.12

This feature actually works better than you would think it would and should work on other keyboards besides the default Google keyboard. As you can see, it worked on my third party Swiftkey keyboard.

Google Automated Assistant "Ok Google"
If you have ever had an iPhone or know somebody that does, I'm sure you have heard of Siri, which is Apple's automated assistant that can be used to do things like check the weather, make a call, check movie times, and so on. To activate Siri, you need to say "hey Siri" and she will come to life, waiting for your next command.

Google has its own version of Siri which is called the Google Assistant. It works the same way as Siri, and can perform the same sort of tasks, but instead of saying "hey Siri" to activate it, you will say "Ok Google." Once it hears the magic words, it will beep at you and be ready to take orders. You will also get a visual notification on your screen that you can interact with (as seen in figure 16.13).

Figure 16.13

You can use the Google Assistant to do things such as get directions, make calls, set reminders, check traffic, and much more. For example, if you want to call Bob Smith's cell phone number, all you need to say is "Ok Google, call Bob Smith mobile" and it will do the rest. If you only have one number for a contact, you can just say "Ok Google, call Bob Smith" since it will dial the only number it has. If you do have more than one number and don't specify which one to call, then you will be prompted as to which number to use.

If your Google Assistant is not working, that probably means it's not activated. To turn it on, try long pressing on your phone's home button and see if it gives you the option to turn it on. If not, then you can try going to the Google app, which you should have installed on your phone, and then find the settings for that app.

From there, find the *Google Assistant* section, and then the *Voice Match* option and turn it on from there.

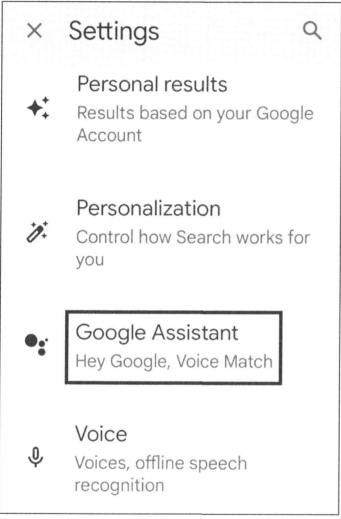

Figure 16.14

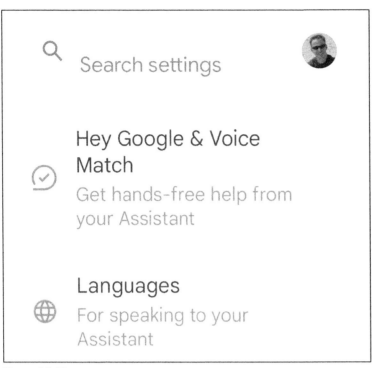

Figure 16.15

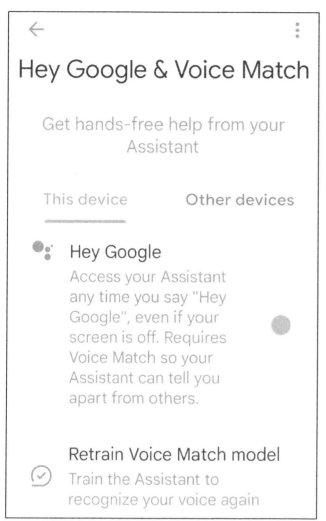

Figure 16.16

You may be prompted to train the Google Assistant to work with your voice, so if that's the case, just follow along with what it asks you to do and you should be all set!

What's Next?

Now that you have read through this book and taken your Android smartphone skills to the next level, you might be wondering what you should do next. Well, that depends on where you want to go. Are you happy with what you have learned, or do you want to further your knowledge on Android devices?

If you do want to expand your knowledge, then you can look for some more advanced books or ones that cover the specific technology that interests you such as Android operating specific book. Focus on one subject at a time, then apply what you have learned to the next subject.

There are many great video resources as well, such as Pluralsight or CBT Nuggets, which offer online subscriptions to training videos of every type imaginable. YouTube is also a great source for instructional videos if you know what to search for.

If you are content in being a proficient smartphone user that knows more than your friends, then just keep on practicing what you have learned and don't be afraid to poke around with new apps, settings, and other configurations because you might be surprised at how well you can make your Android phone work for you and have it do things you didn't think were possible.

Thanks for reading **Android Smartphones Made Easy**. You can also check out the other books in the Made Easy series for additional computer related information and training. You can get more information on my other books on my Computers Made Easy Book Series website.

https://www.madeeasybookseries.com/

What's Next?

What's Next?

You should also check out my computer tips website, as well as follow it on Facebook to find more information on all kinds of computer topics. I also have online training courses you can take at your own pace.

https://www.onlinecomputertips.com
https://www.facebook.com/OnlineComputerTips/
http://madeeasytraining.com

About the Author

James Bernstein has been working with various companies in the IT field for over 20 years, managing technologies such as SAN and NAS storage, VMware, backups, Windows Servers, Active Directory, DNS, DHCP, Networking, Microsoft Office, Exchange, and more.

He has obtained certifications from Microsoft, VMware, CompTIA, ShoreTel, and SNIA, and continues to strive to learn new technologies to further his knowledge on a variety of subjects.

He is also the founder of the website onlinecomputertips.com, which offers its readers valuable information on topics such as Windows, networking, hardware, software, and troubleshooting. James writes much of the content himself and adds new content on a regular basis. The site was started in 2005 and is still going strong today.

Made in the USA
Columbia, SC
11 December 2023

28277692R00143